# Chemical Ecotoxicology

Jaakko Paasivirta

LEWIS PUBLISHERS

**Library of Congress Cataloging-in-Publication Data**

Paasivirta, Jaakko.
Chemical ecotoxicology / Jaakko Paasivirta.
p. cm.
Includes bibliographical references and index.
ISBN 0-87371-366-4
1. Environmental chemistry. 2. Pollutants—Toxicology.
I. Title.
[DNLM: 1. Environmental Pollutants—analysis.
2. Environmental pollutants—toxicity. WA 671 P111c]
TD193.P32 1991
628.5′2—dc20
DNLM/DLC
for Library of Congress 91-17422
CIP



Direct all inquiries to CRC Press, Inc., 2000 Corporate Blvd., N.W., Boca Raton, Florida 33431.

PRINTED IN THE UNITED STATES OF AMERICA
4 5 6 7 8 9 0
Printed on acid-free paper

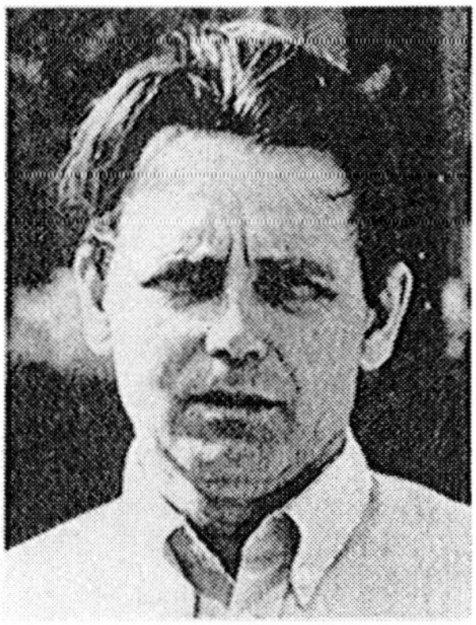

**Jaakko Paasivirta** is professor of organic chemistry at the University of Jyväskylä, Finland. He has received M.Sc. and Lic.Phil. degrees from the University of Helsinki, and a Dr.Phil. degree from the University of Oulu, Finland. In 1966–69 he served as an associate professor at the University of Turku, and from 1970 has served in his present position in Jyväskylä. In 1963–64 he worked as a post-doctoral fellow with Dr. Sidney Brownstein in the Applied Chemistry Division of the National Research Council, Ottawa, Canada. He has been a member of the Finnish Academy of Sciences since 1979.

His research was first directed to structural chemistry and reaction mechanisms of alicyclic compounds and pharmaceutical chemicals and IR and NMR spectroscopy. Since 1970 a continuously increasing part of his activities has been directed to studies on environmental organic pollutants and their model substances and trace analysis methods, such as chromatography/spectroscopy, including the use of biosensors. His present main research projects aim to apply structural organic chemistry to chemical ecotoxicology. These activities include a worldwide cooperation with various research groups. Paasivirta is the author or co-author of more than 430 scientific papers and reports (mainly in English, German, or Finnish) and more than 20 teaching texts in Finnish.

# Contents

# List of Figures

# List of Tables

# 1

# Introduction

Adverse effects of the activities of mankind upon the environment are of rapidly growing concern. Chemical pollution is one of the most pronounced consequences of industrialization and of the intensified application of the uses of natural resources by agriculture, forestry, and mining.

Increased air emissions of dust and inorganic and organic gases form a pronounced threat to the environment and to human health. In addition, increased emissions of toxic metals, and some persistent and toxic organic chemicals, represent a serious hazard to humans and wildlife. Risk assessment of these potential environmental toxicants followed by restrictive and remedial measures (risk management) is essential to preserve the health of the people and the life of nature for the future.

Ecotoxicology aims to discover the chemicals that pose risks in order to apply preventive measures before any serious damage to natural ecosystems occurs. A number of physical, chemical, and biological tests to predict the exposure-related fate and the toxic effects of chemicals in the environment are applied together. These methods, together with chemical studies in the environment, belong to the discipline of chemical ecotoxicology

In other words, the following definition can be given:

*Chemical ecotoxicology applies methods of chemistry to the study of the occurrence and adverse effects of anthropogenic chemicals and their transformation products in the biosphere.*

The effects of chemicals are connected to physical, chemical, and biological processes in the environment. The natural phenomena in which the anthropogenic chemical influences appear can be described by cycles of carbon, oxygen, sulfur, phosphorus, and heavy metals.

Anthropogenic interference with a natural cycle can lead to incidences where a natural compound is detrimental. Good examples are the increases of sulfur, nitrogen, and carbon oxides in the atmosphere and the enrichment of methyl mercury in biota.

The actual anthropogenic chemicals enter into the environment from industrial processes, combustion, and use as food preservatives and additives, packaging materials, textile chemicals, construction and furniture materials, plastics, paints, fuels, fertilizers, pesticides, and hygiene chemicals.

There are about 10 million different synthetic chemicals known. Of these, about 120,000 are in common use and about 11,000 are produced in amounts higher than 500 kg per year. Of the total material in the environment, only 1/1000 is anthropogenic, but this portion includes some toxic and very persistent compounds that accumulate in living things to harmful levels. These compounds are the environmental toxicants.

# 2

# Cycles of Chemicals in the Environment

## 2.1 COMPARTMENTS

### 2.1.1 Definition

Our environment can be compartmentalized as follows: lithosphere (solid outer mantle of the earth), hydrosphere (water), atmosphere, biosphere (all living things), and exosphere (outer space). For chemical ecotoxicology, the first four of these compartments are important. Their names and abbreviations as used in this book are soil (s), water (w), air (a), and biota (b).

The occurrence and fate of chemicals in the environment includes transfers between the compartments (e.g., by evaporation or precipitation) and inside them (from one location to another or from one living individual to another). Hazardous chemicals move fastest and widest in air and water; therefore, these two compartments are of greatest interest to international environmental protection.

### 2.1.2 Soil

The lithosphere consists of the solid outer mantle of the earth. It is exposed to the atmosphere over one quarter of the earth's surface area. The upper layer of this portion of the lithosphere is widely covered by soil. Other parts of the earth's crust are covered with water, and generally the uppermost layer consists of sediment.

Parent rocks of the soil are disintegrated by physical weathering into rock fragments and minerals, a process that further releases nutrients for plants. Residues from organisms return to the weathering mass to be transformed into humus. Soils are washed by water to rivers, lakes, and further to oceans, where their residues are deposited and eventually give rise to rock formation.

A living surface soil is actually a mixture of solid matter, water, and biota components. The fate of a chemical in soil depends on its sorption potency to solid particles, its water and lipid solubility, and its persistence against hydrolytic, catalytic, and biological factors. High organic-matter (humus) content in soil leads, in general, to higher persistency and lower mobility of organic chemicals in soil. Therefore, in Nordic countries where the humic content is high and temperatures low, many agricultural pesticides may remain in soil for longer periods than desirable and thus cause toxic problems. On the other hand, groundwater in these countries is less vulnerable due to the slow leaching of organic chemicals in highly humic soils.

### 2.1.3 Water

The molecular structure of water, in which hydrogen atoms are bound to oxygen with an H–O–H angle of 104.5°, explains its dipole character and strong hydrogen bonding to other water molecules in three dimensions. Consequently, water has a large heat capacity and an extremely high heat of vaporization. Water also has a density maximum at 4°C. These physical properties characterize it as the central substance that supports the present forms of life.

Global water originates from very long-term reactions of volcanic gases and carbon dioxide (acids) with rocks (oxides, carbonates, and silicates). The presently existing water stays in dynamic equilibrium between different reservoirs in the hydrologic cycle (Figure 2.1.1).

The sweetwater originates from precipitation, i.e., rain or snowfall. It forms the transport medium and solvent in the reactions and processes in which the solid rock degrades and moves to dissolved matter, soil, sediments, and sedimentary rocks. Surface waters and water in the atmosphere compose a very small fraction of the total water of the earth, but fluxes into and out of these reservoirs are relatively high and the residence time short. The groundwater reservoir is much greater, but it is less mobile and therefore less significant in transport processes than the atmospheric and surface waters. The largest sweetwater reservoir is ice, where 3% of all water is kept in a relatively immobile stage.

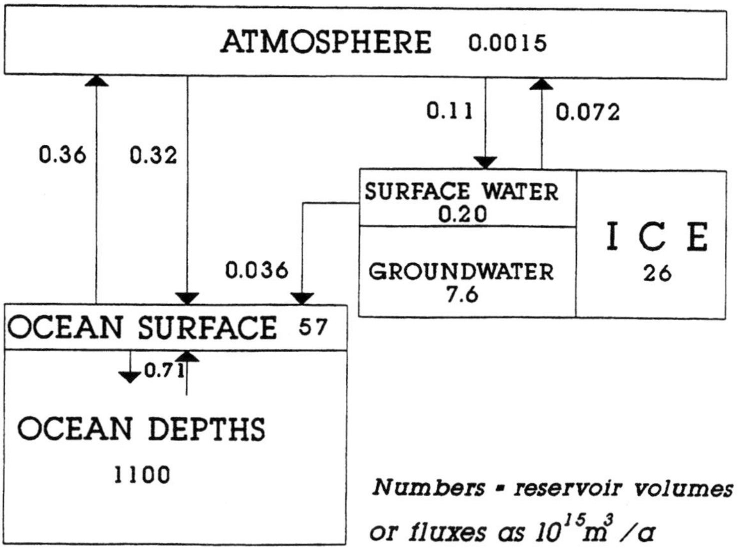


**Figure 2.1.1** The hydrologic cycle. (From *The Handbook of Environmental Chemistry, Vol. 1, Part A,* Hutzinger, O., Ed., Springer-Verlag, Berlin, 1980. With permission.)

The major reservoir (97%) of water is the oceans. About 5% of this water is in the upper mixed layer, where most of the biological activity occurs. The remaining ocean water is in the deep ocean, with residence time of approximately 1600 years. In contrast, the residence time of water in upper layer of the oceans is approximately 80 years, and in the atmosphere only 1.3 days.

## 2.1.4 Air

The atmosphere is a layer of gases surrounding the earth with decreasing density up to 500 km above the earth's surface. The atmosphere regulates surface conditions that are essential to life. The two most important regulatory processes are absorption of ultraviolet solar radiation and generation of the greenhouse effect.

Chemical composition of the air (Table 2.1.1) is homogenous up to 100 km above the earth's surface. The bottom 0 to 15 km layer, the troposphere, contains 80% of the total mass of the atmosphere.

The most abundant nonvariable gases (nitrogen, oxygen, and argon) account for more than 99.9% of the total atmospheric

**Table 2.1.1 Composition of the atmosphere in 1970**[1]

| Constituent | Formula | Abundance by Volume | Constituent | Formula | Abundance by Volume |
|---|---|---|---|---|---|
| Nitrogen | $N_2$ | 78.084% | Methane | $CH_4$ | 2 ppm |
| Oxygen | $O_2$ | 20.984% | Hydrogen | $H_2$ | 0.5 ppm |
| Argon | Ar | 0.934% | Nitrous oxide | $N_2O$ | 0.3 ppm |
| Water vapor | $H_2O$ | Variable | Carbon monoxide | CO | 0.05–0.2 ppm |
| Carbon dioxide | $CO_2$ | 325 ppm | Ozone | $O_3$ | Variable |
| Neon | Ne | 18 ppm | Ammonia | $NH_3$ | 4 ppb |
| Helium | He | 5 ppm | Nitrogen dioxide | $NO_2$ | 1 ppb |
| Krypton | Kr | 1 ppm | Sulfur dioxide | $SO_2$ | 1 ppb |
| Xenon | Xe | 0.08 ppm | Hydrogen sulfide | $H_2S$ | 0.05 ppb |

composition. Water vapor is a rather important component whose concentration is highest in tropical and lowest in cold continental air masses of the circumpolar regions. Carbon dioxide content has varied considerably through the millennia. At present, it is increasing 0.7 to 1 ppm/a due to anthropogenic influences, such as fossil-fuel burning.

In upper parts of the troposphere, water vapor, carbon dioxide, methane, nitrogen oxides, and some anthropogenic gases absorb infrared radiation, which is formed from solar radiation reflecting from the earth's surface. This phenomenon, called the greenhouse effect, partially prevents solar energy from escaping back into space. It is an important regulator of the earth's climatic conditions. All parts of the solar radiation below 320 nm wavelength are absorbed by gases in layers above the troposphere (0 to 11 km from surface). These layers are the stratosphere (11 to 50 km), the mesosphere (50 to 85 km), and the thermosphere (85 to 500 km). The shortest wavelengths below 234 nm break the bond in the oxygen molecules, which leads to the formation of ozone (reaction (1) below):

$$O_2 + hv(\lambda < 234\ \text{nm}) = O^{\cdot} + O^{\cdot} \xrightarrow{+2O_2} 2O_3 \quad (1)$$

$$O_3 + hv(\lambda < 320\ \text{nm}) = O^{\cdot} + O_2 \quad (2)$$

$$O_2 + O^{\cdot} = O_3 + \text{Energy} \quad (3)$$

$$O_3 + O^{\cdot} = 2O_2 \quad (4)$$

The UV radiation of 234 to 320 nm is absorbed by ozone in the bond-breaking reaction (2) to form oxygen and an oxygen atom (radical), which combine again to form ozone in reaction (3). The ozone chain (2) to (3) is terminated by reaction (4), in which an ozone molecule combines with an oxygen radical.

The ozone-oxygen photoreaction chain contributes to the increased temperatures of gases in the stratosphere. In the uppermost portion of the troposphere (tropopause), the temperature is 217 K, but in the upper part of stratosphere, the maximum temperature of 270 K is reached. In the mesosphere, fewer exothermic reactions occur, and its temperature decreases with increasing height down to 100 K in the mesopause (85 km) layer.

The shortest wavelength ($\lambda < 234$ nm) region of solar irradiation is absorbed in the thermosphere in photoionization reactions of oxygen and nitrogen molecules and oxygen atoms. The energy formed heats the gases to 1000 to 2000 K. The photoionizations at a height of 300 to 500 km are sources of the polar lights (Aurora Borealis).

## 2.2 THE CARBON CYCLES

### 2.2.1 Introduction

Carbon is transported between the atmosphere, hydrosphere, soil, and biota in a short-term cycle having a time scale of tens to hundreds of years. But part of the carbon is removed by sedimentation from this biocycle to the long-term geochemical carbon cycle, which has a million-year time scale.

### 2.2.2 Geochemical Carbon Cycle[2]

The long-term movements of carbon take place in the carbonate-silicate geochemical cycle, where carbon dioxide is removed from the atmosphere, stored in carbonate rocks, and then returned into the air.

Carbonates form when carbon dioxide dissolves in rain and reacts chemically with rocks that contain calcium-silicate minerals. Such reactions release calcium and bicarbonate ions ($Ca^{2+}$ and $HCO_3^-$) into groundwater, which transports the ions to streams, rivers, and finally the ocean. In the ocean, plankton and other organisms use the ions to construct shells of calcium carbonate ($CaCO_3$). The shells are deposited in sediments on the sea floor when the organisms die.

Slowly the sea floor spreads; eventually it and associated sediments slip under the continents. Exposed to increased temperature and pressure, the sediment releases carbon dioxide gas, which then reenters the atmosphere via volcanic emissions.

Before the onset of life, the deposition of calcium carbonate in the sea took place after concentrations of calcium and carbonate ions reached the critical level: the cycle is operable also without the participation of the biota.

Increased silicate weathering decreases the carbon dioxide content of the atmosphere. This process took place in periods after the formation of new mountains. Consequently, the greenhouse effect decreased and a cold climatic period with glaciers followed. This should be the natural situation at present, but anthropogenic influences counteract and perhaps reverse this effect by increasing the atmospheric carbon dioxide levels.

Increase of the global temperature causes processes that increase the binding of carbon dioxide to rocks. Therefore, the carbonate-silicate cycle creates the temperate climate of earth. However, this natural regulation works in periods of million years; the anthropogenic increase of atmospheric carbon dioxide might cause serious short-term climatic changes before the long-term cycle balances the conditions.

### 2.2.3 Biocycle of Carbon

The carbon biocycle has two separate rotations, one on soil-covered and the other on the sea-covered areas (Figure 2.2.1).

Annually, soil plants bind 35 billion and sea phytoplankton 40 billion tons of carbon from the carbon dioxide in the atmosphere and water. Terrestrial animals, zooplankton, and fish consume the bound carbon and combine it again with oxygen either in air or dissolved in water. This respiration process, together with the degradation of dead organic matter, returns carbon dioxide to be available for plants. Anthropogenic combustion of the fossil fuels at level of 5 billion tons/a, as it was in 1980, liberates so much $CO_2$ that the annual increase of its atmospheric content should be 2 ppm. The measured increase, however, was 0.7 ppm/a in 1980. This is explained by the buffering uptake of $CO_2$ by biota.

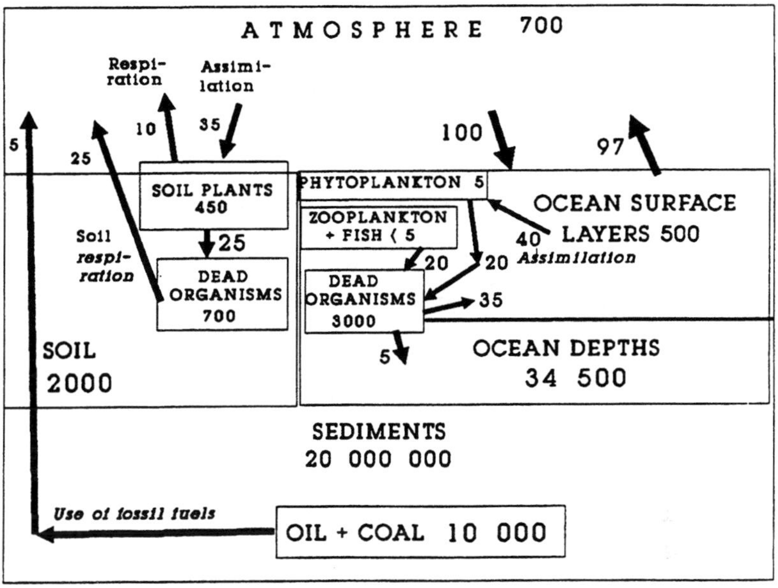


**Figure 2.2.1** Reservoirs and annual fluxes of carbon in billions of metric tons ($\times 10^9$ t).

The content of carbon dioxide in different parts of the air at different times shows the importance and speed of the assimilation and respiration processes in the carbon biocycle. For example, when a measured minimum at noon in a dense forest was 305 ppm, it increased at night to 350 ppm just before dawn. Monthly averages at a several-kilometer height (e.g., Mauna Loa station, Hawaii) vary between the summer minimum and winter maximum by approximately 6 ppm.

## 2.2.4 Carbon Dioxide and the Increasing Greenhouse Effect

Annual averages of the atmospheric carbon dioxide are increasing at present about 1 ppm/a. If this trend continues, the greenhouse effect of $CO_2$ will cause an increase in the average global temperature. Consequent changes in weather patterns could cause serious problems, such as loss of productivity in large agricultural areas and the melting of ice caps. The latter phenomenon would raise the sea surface enough to submerge some major cities.[3]

The contribution of water vapor in air to changes in the

greenhouse effect is very difficult to predict, because reflection from the clouds compensates for the adsorption of infrared radiation by gaseous water molecules. In contrast, anthropogenic increase of small molecules such as nitrogen oxides, methane, and halohydrocarbons in the troposphere will elevate the development of the greenhouse warming of earth.

## 2.3 THE OXYGEN CYCLE

The atmosphere of the earth, rich in free-oxygen gas, is a great exception in the known universe. Elsewhere, oxygen is a rather rare element and exists primarily bound to other elements. The observable sources of the free oxygen gas are the assimilating plants: in fact photosynthesis is the only significant producer of elemental oxygen in our environment.[1]

Excess oxygen in the atmosphere is returned to its compounds by respiration, by combustion, and (more slowly, but significantly) by reactions with the reduced minerals of soil. The biogeochemical oxygen cycle is described in Figure 2.3.1.

Annual gains in atmospheric oxygen (in $10^{12}$ moles/a) arise through four processes:[1]

| | |
|---|---|
| 1. Net photosynthesis on land | 4600 |
| 2. Net photosynthesis in the ocean | 3790 |
| 3. Photolysis of $N_2O$ | 0.36 |
| 4. Escape of hydrogen to space | 0.007 |

The losses of atmospheric oxygen (in $10^{12}$ moles/a) arise through 19 different processes, of which only two (18 and 19) are anthropogenic:[1]

| | |
|---|---|
| 1. Aerobic respiration | 6730 |
| 2. Photochemical oxidation of $NH_3$ | 4.3 |
| 3. Microbial oxidation of $NH_3$ (nitrification) | 1200 |
| 4. Photochemical oxidation of $CH_4$ | 113 |
| 5. Microbial oxidation of $CH_4$ | 226 |
| 6. Photochemical oxidation of terpenes | 45.5 |
| 7. Photochemical oxidation of $H_2S$ | 1 |
| 8. Microbial oxidation of $H_2S$ | 2 |
| 9. Photochemical oxidation of $CH_3SCH_3$ | 2.5 |
| 10. Surface reaction of $O_3$ | 24 |

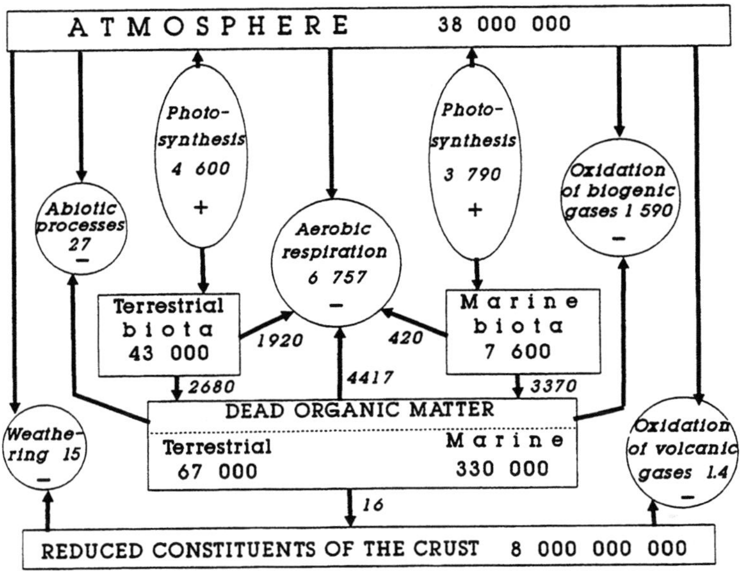


**Figure 2.3.1** The oxygen cycle. Reservoirs, fluxes, and annual gains (+) or losses (–) of atmospheric oxygen. Numbers equal amounts in $10^{12}$ moles of oxygen or the capacity to combine with oxygen.[1] (From *The Handbook of Environmental Chemistry, Vol. 1, Part A*, Hutzinger, O., Ed., Springer-Verlag, Berlin, 1980. With permission.)

| | |
|---|---|
| 11. Fixation of nitrogen by lightning | 3.3 |
| 12. Oxidation of volcanic $H_2$ | 0.55 |
| 13. Oxidation of volcanic CO | 0.15 |
| 14. Oxidation of volcanic $SO_2$ | 0.7 |
| 15. Chemical weathering of kerogen carbon | 7.5 |
| 16. Chemical weathering of sulfide sulfur | 3.8 |
| 17. Chemical weathering of ferrous ion | 1.3 |
| 18. Combustion of fossil fuel | 350 |
| 19. Industrial fixation of nitrogen | 2.7 |

Degradation of the weathered organic carbon (kerogen) is aerobic at the surface layer and anaerobic (fermentation to methane) at deeper layers of the sediments. These processes are very slow, and their rates decrease with increasing depth of the sediment layers. Therefore, a significant part of kerogen returns to the sedimentary rocks after one weathering/sedimentation cycle. The weathered sulfides react much faster, and their leaching/deposition equilibrium might have a more significant role in the development of the atmospheric oxygen content.

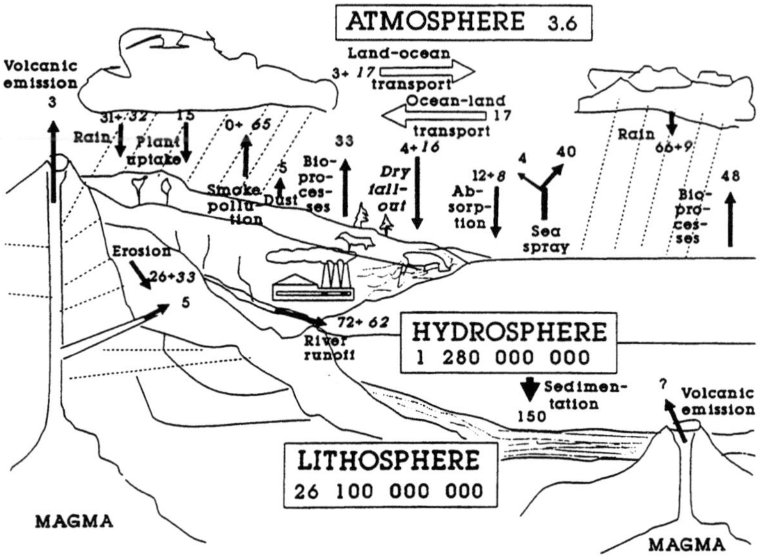


**Figure 2.4.1** The global sulfur cycle. Reservoirs and annual fluxes are shown in millions of tons. On the flux arrows, natural transfers are indicated in roman, anthropogenic influences in italic numbers. Flux from volcanoes in bottoms of the oceans is unknown. (From *The Handbook of Environmental Chemistry, Vol. 1, Part A,* Hutzinger, O., Ed., Springer-Verlag, Berlin, 1980. With permission.)

## 2.4 THE SULFUR CYCLE

The biosphere has a large influence on the global sulfur cycle, which is complex due to the occurrence of many oxidation stages of sulfur between –2 and +6. The main processes of the global sulfur cycle are illustrated in Figure 2.4.1.

The main portion of the moving sulfur is in the form of $SO_2$ and $SO_4^{2-}$. Reduced sulfur ($H_2S$) appears as a fraction of the volcanic emissions and biological degradation processes of the organic sulfur compounds. The residence time of sulfur in the atmosphere is only 5.7 days, but in oceans it is 40 million years and in sediments and sedimentary rocks 9.7 billion years. The main anthropogenic input to the sulfur cycle is the combustion of fossil fuels, which was estimated in the 1960s to be 64 to 65 million tons per year.[1] This emission causes acid rain, which lowers the pH of lake waters and mobilizes toxic metals such as aluminum, cadmium, lead, and mercury. Consequently, lake fauna and forest flora are damaged at many areas to a disastrous degree.

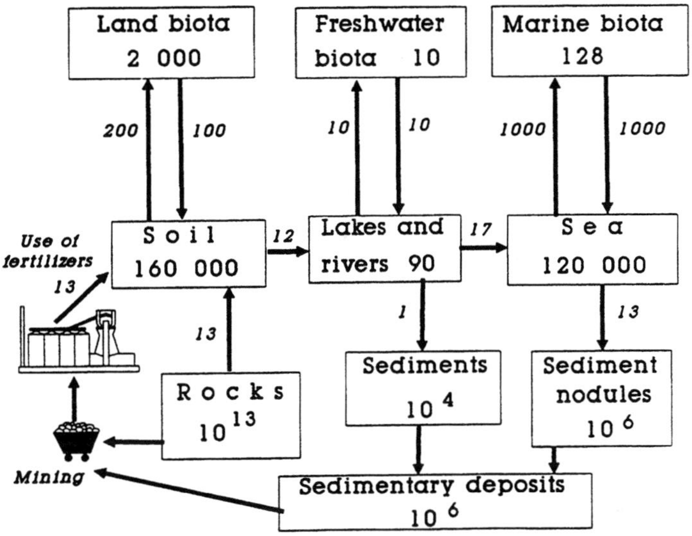


**Figure 2.5.1** Reservoirs and annual fluxes of the global phosphorus in millions of tons (Mt). One Mt of elemental P corresponds to five Mt of fluoroapatite mineral. (From *The Handbook of Environmental Chemistry, Vol. 1, Part A,* Hutzinger, O., Ed., Springer-Verlag, Berlin, 1980. With permission.)

## 2.5 THE PHOSPHORUS CYCLE

Phosphorus is a vitally important element for life and essential for nutrition of the biosynthetic processes of the organisms. The chemistry of phosphorus is very rich in different compound types. Paradoxically, important compounds in natural cycling and life are almost solely compounds of $P^V$, salts and organic esters of phosphoric acid. A vast number of catalysts and reagents of physiological processes are organic derivatives of phosphoric acid. The mass balance of the phosphorus cycle is shown in Figure 2.5.1.

Anthropogenic influences upon the phosphorus cycle primarily cause local aquatic ecosystem disturbances due to eutrophication by excess phosphorus from fertilizers. The consequences are, for example, a decrease in water quality, blooming of toxic algae, and loss of edible fish species.

## 2.6 THE HEAVY METAL CYCLES

### 2.6.1 Introduction

Heavy metals are naturally mobile between different components of the earth's environment. Anthropogenic activities re-

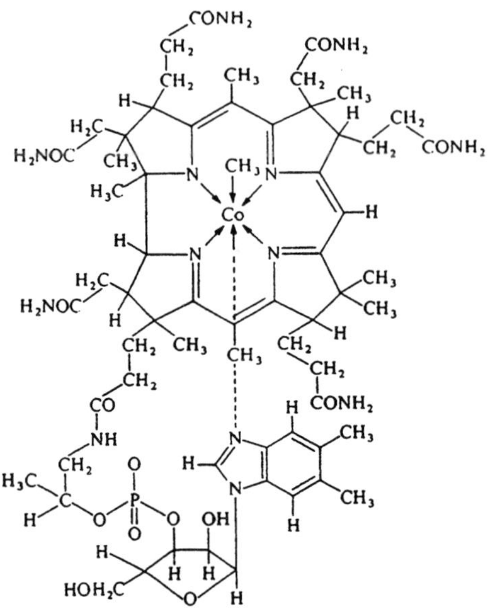

**Figure 2.6.1** The structure of methyl cobalamin ($CH_3CoB_{12}$). (From *The Handbook of Environmental Chemistry, Vol. 1, Part A,* Hutzinger, O., Ed., Springer-Verlag, Berlin, 1980. With permission.)

sult in minor changes in the equilibria of these cycles on a global scale. However, at some localities man's contribution can be significant. The most hazardous heavy metals are the toxic and industrially important mercury, lead, tin, and cadmium.[1]

## 2.6.2 Biomethylation

Biomethylation is the stage in the biogeochemical cycle that elevates the ecotoxicity of mercury, lead, and tin. There are several bioreactions in microorganisms that methylate inorganic cations. One of the best known involves methyl cobalamin (a modification of vitamin $B_{12}$; see Figure 2.6.1) as a donor of carbanion $CH_3^-$. Similar carbanion transfers to metal cations are mediated in biota by s-adenosyl-methionine and by $N^5$-methyltetrahydrofolate derivatives.[1]

## 2.6.3 Metal Cycles

A generalized scheme of biogeochemical cycling of a metal M is shown in Figure 2.6.2.

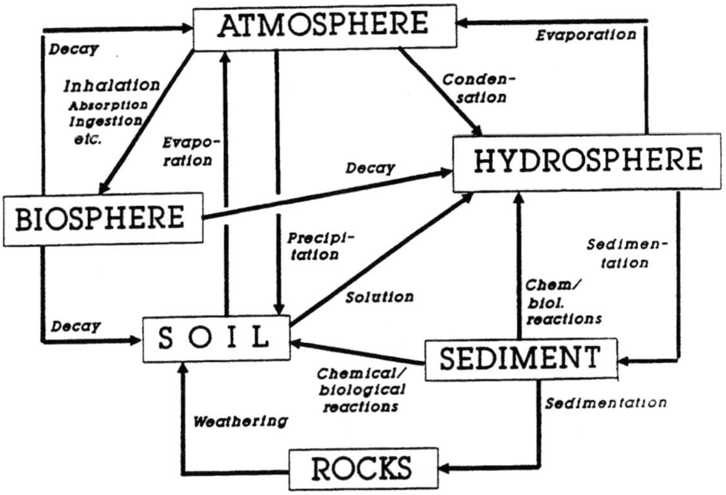


**Figure 2.6.2** Main transport processes of the biogeochemical cycle of a metal M that occurs in the environment both as cations and as organic derivatives.

## 2.6.4 Mercury Cycle

Biomethylation is a very important process in the mercury cycle. In the soil and in surface layer of sediment, microbia transfer $Hg^{2+}$ to monomethyl mercury chloride. This compound is lipophilic and accumulates in biota. In the atmosphere, mercury occurs primarily in its most volatile forms: elemental vapor, methyl mercury, and dimethyl mercury. In the hydrosphere-atmosphere system, the latter two are in equilibrium with dissolved methyl mercury:

$$HgCl_2 \rightleftharpoons H_3C-HgCl \rightleftharpoons H_3C-Hg-CH_3$$

More details of the natural mercury cycle and anthropogenic interferences are described in Chapter 7.

## 2.6.5 Lead Cycle

Natural weathering moves 11,000 tons of lead (Pb) annually to the oceans. The anthropogenic addition from agriculture and land constructions to this is 33,000 t/a. The mining industry produces 4 million tons of lead annually. The effluent of lead from industry to waters is 140,000 and from car traffic 300,000 tons per year (in 1970).[1] The origin of the latter emission is tetraethyl lead, an acute toxicant, used in gasoline. New installations of exhaust catalytes in cars removes this problem, because unleaded gasoline is then used.

Both organic and inorganic lead can accumulate in animal and human tissues to toxic levels.

Lead biomethylates in the environment to tetramethyl lead (TML) which is bioavailable to such a degree that 10 to 24% of the total lead content in fish muscles consists of TML. Compared to methyl mercury (96 to 100% of the total mercury in fish muscles), TML has less importance as an environmental hazard.

### 2.6.6 Tin in the Environment

Tin (Sn) occurs in nature as oxide ores and as organic complexes in peat and coal. It is a widely distributed trace element. Contents in surface waters in the U.S.A. are 0.8 to 30 μg/l (ppb), in sea animals 0.2 to 20 ppm, and in terrestrial animals generally lower than 0.15 ppm. The residence time of tin in the oceans is 100,000 years.[1]

Production of tin by mining is 200,000 tons per year. Its main uses are plating (50%), soldering (25%), bronze (7%), babbitt metal (3%), tinning (2%), and other uses (12%). Dust at industrial areas contains up to 10 mg/g Sn.

About 25,000 tons of tin(IV)-organic compounds are produced annually for use as stabilizers (e.g., in polyvinyl chloride) and biocides including disinfectants, pesticides, and marine antifouling paints. These substances are bioaccumulating, persistent, and toxic, and appear as ecological problems in many localities. Inorganic tin is also biomethylated in the environment.[1]

### 2.6.7 Cadmium

Cadmium (Cd) occurs mainly at very low levels in most compartments of the ecosphere. Its natural cycle is of no toxic concern, but human activities cause Cd problems in many localities. The measured contents and residence times for cadmium are as follows:

| | | |
|---|---|---|
| In air | 0.1—500 $ng/m^3$ | 20—30 days |
| In water | 0.01—42,000 μg/l | 1—2 years (lake)<br>250,000 years (ocean) |
| In sediments | 0.01—50,000 μg/g | 2—5·$10^8$ years |
| In organisms | 0.001—1120 μg/g | |

Cadmium is mainly recovered as a by-product of zinc metallurgy. Annual production is 15,000 to 20,000 tons. It is used in electroplating (34% of total consumption), in nickel-cadmium batteries (15%), as a polyvinylchloride (PVC) stabilizer (15%), in solders, brazing materials, and other alloys (8%), and in many other technologies. Discharges and combustion of wastes causes Cd emissions that create local and widespread contamination.[4]

Cadmium is one of the most toxic elements. It is not biomethylated to any significant extent, but it is bioaccumulating. Mobilization of Cd takes place through chelation with organic hydroxy acids. In animals, cadmium binds to metallothionein and other sulfur-containing proteins. Especially high levels of Cd have been found in the hepatopancreas of lobsters living in cadmium-contaminated coastal areas of the sea.[5,6] From contaminated soil it is taken up by plants and then enriched in kidneys and livers of the herbivore animals.[4]

Industrial cadmium discharge caused a fatal epidemic of poisonings in the Jintsu River area, Japan 1947 (100 deaths until end of 1965). The symptoms were extremely painful: the disease was called *itai-itai* (meaning "ouch-ouch").

## 2.7 OTHER METALS[1]

Aluminum, iron, zinc, chromium, and copper are common metallic elements that are strongly bioaccumulating in their salt form. Their increased release from the soil due to acid precipitation causes toxic effects to plants and to aquatic organisms in lakes.[18]

Dissolved aluminum in its various inorganic and organic compounds is both a human neurotoxicant and a serious threat to plants and wildlife.[7] Iron, zinc, chromium, and copper cause local toxic effects to biota through high concentrations caused by heavy industrial discharges. Their global threat as a possible ecological hazard is small related to Hg, Pb, Sn, Cd, and dissolved Al described above.

## 2.8 SUMMARY

Element cycles describe how chemicals move between the soil, water, air, and biota compartments of the ecosphere. Physical, chemical, and biological processes in each compartment influence the cycling and the movements and transformations of the compounds within that compartment. The fastest cycling processes take place in air and water; therefore, pollution of these compartments has a global significance. The increase of carbon dioxide and organohalogen compounds in the atmosphere due to human activities might lead to ecological damage through influences on the greenhouse effect and on the effect of solar radiation upon the earth's surface. Other anthropogenic ecotoxic effects can be described in the cycles of carbon, oxygen, sulfur, phosphorus, and metals.

## 2.9 REFERENCES

1. **Hutzinger, O., Ed.** *The Handbook of Environmental Chemistry, Vol. 1, Part A* (Berlin: Springer, 1980).
2. **Kasting, J.F., O.B. Toon, and J.B. Pollack.** "How Climate Evolved on the Terrestrial Planets," *Sci. Am.* 258 (2):45-53 (1988).
3. **Cooper, D.** "Modelling the Greenhouse Effect," *Educ. Chem.* May, 75-77 (1989).
4. **Förstner, U.** "Cadmium," in *Anthropogenic Compounds. The Handbook of Environmental Chemistry, Vol. 3, Part A,* O.Hutzinger, Ed. (Berlin:Springer, 1980), pp. 59-107.
5. **Ray, S., and M. White.** "Methallothionein-like Protein in Lobsters," *Chemosphere* 10:1205-1213 (1981).
6. **Ray, S., D.W. McLeese, and L.E. Burridge.** "Cadmium in Tissues of Lobsters Captured Near a Lead Smelter," *Mar. Pollut. Bull.* 12:383-386 (1981).
7. **Lewis, T.E.** *Environmental Chemistry and Toxicology of Aluminium* (Chelsea, MI:Lewis Publishers, 1987).

# 3

# Risk Assessment and Management

## 3.1 NATURE OF A RISK

Risk is the likelihood of the realization of a hazard. Risk can also be defined as the expected magnitude (probability) of harm associated with an act.[1] It can be *attributable risk,* which is the increase in risk attributable to a particular exposure. The term *relative risk* is used to describe the ratio of the risk to individuals exposed to an agent to the risk to individuals not exposed.[2]

While risk exists only in our imagination as a projection of what might happen, it cannot be measured with instruments. In contrast, a risk situation can be identified and studied objectively.[3]

A risk-situation scheme for a chemical in the environment (Figure 3.1.1) shows that the hazard in this case is the chemical itself. Use of the chemical causes exposure of the vulnerable entities, which might get damaged through primary reactions. The consequences might be death or disease of individuals (like cancer of humans), decline in populations, modification of the ecological community, or disturbances of the ecosystems.

Risk is characterized qualitatively as the nature of the adverse consequences and quantitatively as their probability. It can be seen from the situation scheme (Figure 3.1.1) that the most important factors of the risk are amount of chemical at the target (= exposure) and potency of the chemical to cause the damage (= toxicity). Consequently, quantitative risks may be calculated:

$$\text{RISK} = \text{EXPOSURE} \times \text{TOXICITY}$$

or

$$\text{RISK} = \text{EXPOSURE} \times \text{EFFECT}$$

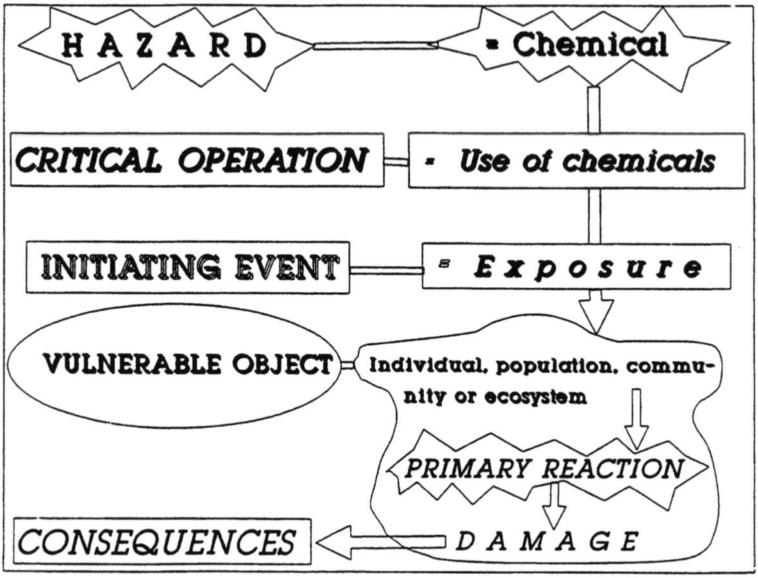


**Figure 3.1.1** A risk situation with a chemical.

## 3.2 RISK MANAGEMENT OF ENVIRONMENTAL CHEMICALS

For the risk assessment of an environmental chemical, the four-step process developed by the U.S. Government can be applied:[4,5]

| | |
|---|---|
| 1. Hazard identification | "Is the chemical toxic?" |
| 2. Dose-response assessment | "How toxic is the chemical?" |
| 3. Exposure assessment | "Who is exposed to the chemical, to how much, for how long and how often?" |
| 4. Risk characterization | "So what? What are the consequences? Do we have a problem?" |

After risk characterization, the acceptable level of exposure via food, water, ambient air, etc. is determined. Alternatives to control the risk are then studied and managers decide how to reduce the risk to acceptable level and execute all necessary

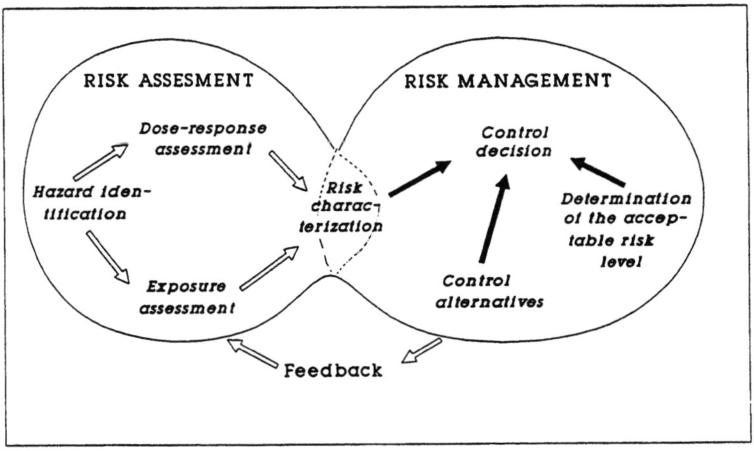


**Figure 3.2.1** The risk management process.

control acts. During this management process, feedback to exposure estimation is done, e.g., by environmental monitoring studies (Figure 3.2.1).

An important example of a risk situation to be managed is environmental pollution.[6] A flowchart for the analysis to find the most effective countermeasures to this central problem of modern societies is described in Figure 3.2.2.

## 3.3 STEPWISE PROCEDURE FOR CONTROL OF CHEMICALS

Properties of pollutants like sulfur and nitrogen oxides in air are well known, and countermeasures can be directed to reduction of their emissions. In contrast, for chemicals that cause problems at low levels, application of chemical and biological techniques is needed to assess actual exposures and possible harmful effects.

The number of old and new chemicals released into the environment by different usages, emissions, and waste treatments is enormous. Rational management of their risk is possible only through a system whereby the fastest and most economical methods are first used to eliminate the majority of harmless chemicals. Only those chemicals that give indications of hazard are further studied in the more laborious steps (Figure 3.3.1).

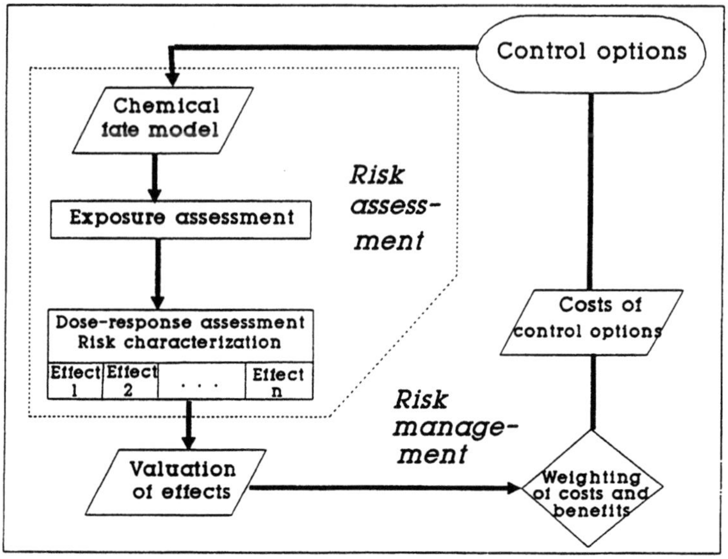


**Figure 3.2.2** Components of an analysis of pollution control.[6] (From *Risk Assessment of Chemicals in the Environment*, Richardsson, M.L., Ed., The Royal Society of Chemistry, London, 1988. With permission.)

Usage patterns, emissions, possible environmental fate data, and modeling using selected properties should be considered first. If the chemical appears to be a potential hazard (+ in Figure 3.3.1), laboratory studies are performed in a sequence, starting with some simple bioaccumulation tests. A hazard (+) indication leads to further testing, but a negative (–) result ends testing for that chemical. In any case, monitoring of the environmental fate of the chemical or of its transformation products should be done to ensure the proper risk management.

## 3.4 PREDICTION OF ENVIRONMENTAL FATE

### 3.4.1 General Principles

Exposure of biota to an environmental chemical is measurable in each compartment of the ecosphere by analyzing samples of the target biota. For risk assessment, however, prediction of the exposures in probable situations of pollution is needed and cannot wait until actual observations are available.

For a new chemical, the first prediction of exposure can be done using a few of its physical and chemical properties and the biodegradability of the compound. These properties can be

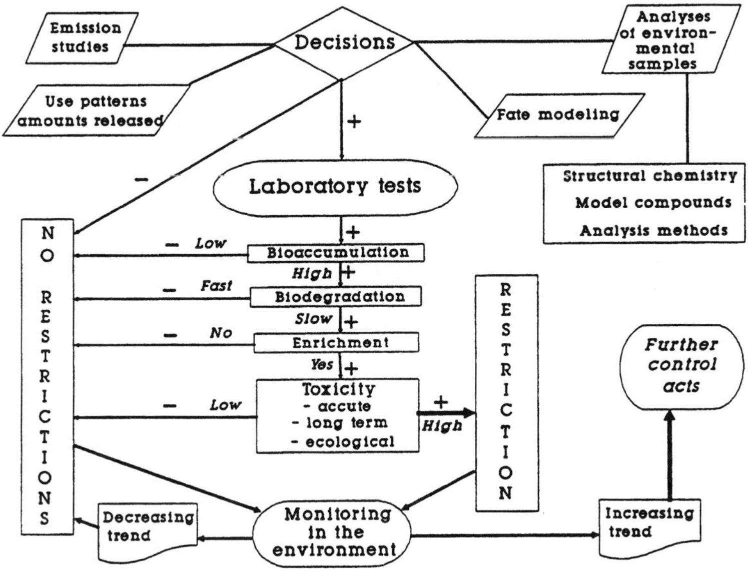


**Figure 3.3.1** Stepwise procedure for control of chemicals against their potential environmental risk.

measured by standard procedures of the chemical laboratory supplemented by some standard tests developed for environmental exposure assessment.[7,8] The property data are used in chemical-fate model calculation, which produces exposure-related values for different model environments and ecosphere compartments.

## 3.4.2 Chemical Fate Modeling

A stepwise approach using unit world model allows prediction of the environmental fate of a new chemical by simple calculations.[9,10]

Compartmentalization is used as the first step. It gives the distribution of the compound between different compartments of the unit world (Figure 3.4.1).

The unit world is a "slice of the earth" with a horizontal area of one km$^2$. Six main compartments are taken into account:

| Compartment | ($i$) | $V_i$ (Volume m$^3$) | $D_i$ (Density kg/m$^3$) |
|---|---|---|---|
| Air | (a) | 6,000,000,000 | 1.22 |
| Water | (w) | 7,000,000 | 1000 |

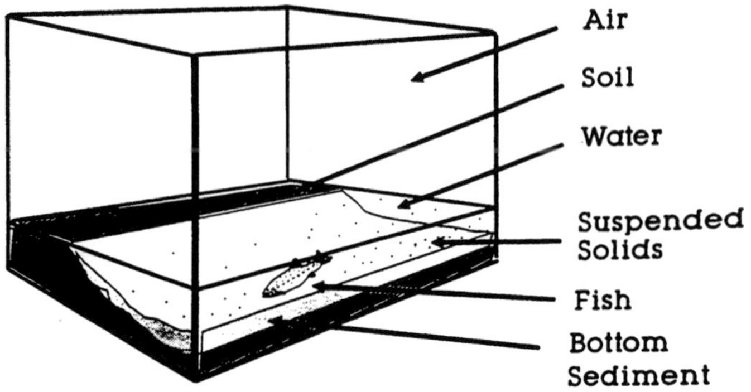


**Figure 3.4.1** Schematic representation of the unit world.[9,10]

| Compartment | (*i*) | $V_i$ (Volume m$^3$) | $D_i$ (Density kg/m$^3$) |
|---|---|---|---|
| Soil | (s) | 45,000 | 1500 |
| Bottom sediment | (bs) | 21,000 | 1500 |
| Suspended solids | (ss) | 35 | 1500 |
| Biota (fish) | (f) | 7 | 1000 |

Level I partitioning calculation is the simplest compartmentalization procedure.[9] There a constant amount of the chemical is added to the unit world and its equilibrium distribution between compartments is calculated. The basis of the calculation is fugacity. Fugacity is related to the partial pressure of the chemical and is a measure of the escaping tendency of the molecule from a phase. At equilibrium, the fugacities are equal in all phases (compartments).

Equilibrium is defined as a series of partition coefficients:

$$H = C_a / C_w, \quad K_p = C_s / C_w, \quad \text{and} \quad \text{Bcf} = C_f / C_w,$$

$H$ = Henry's constant, $C_a$ = concentration in air, $C_w$ = concentration in water, $C_s$ concentration in soil, $C_f$ = concentration in fish, Bcf = bioconcentration factor, and $K_p$ = soil sorption coefficient.

Semiempirical rules are used for estimation of constants that could not be directly measured, e.g., Henry's constant, by the equation:

$$H = (P * M * 16.04) / (T * S),$$

where $P$ = vapor pressure (mmHg), $M$ = molecular weight, $T$ = temperature (K), and $S$ = solubility (g/m$^3$ or ppm).

The soil adsorption coefficient is:

$$K_p = 0.6 * K_{ow} * \mathrm{Ocf},$$

where $K_{ow}$ is the octanol-water partition coefficient and Ocf is the fraction of organic carbon in soil.

Bioconcentration factor can be measured or estimated from $K_{ow}$:

$$\mathrm{Bcf} = 0.048 * K_{ow}$$

If also $K_{ow}$ is unknown, it can be estimated by

$$\log K_{ow} = 6.5 - 0.89 * (\log(S / M)) - 0.015 * \mathrm{Mp},$$

where Mp is the melting point in °C.

The total amount of compound (moles) in the unit world may be expressed as

$$M = \sum V_i C_i$$

and in expanded form,

$$M = C_w \left[ V_a * H + V_w + \left( V_{bs} * K_p + V_{ss} * K_p + V_s * K_p \right) * D_s + V_f * \mathrm{Bcf} * D_f \right] \tag{3.1}$$

Using the above equations, concentrations $C_s$ ($C_s + C_{bs} + C_{ss}$), $C_w$, $C_a$, and $C_f$ can be estimated. The results may be shown in normalized form by setting $C_s + C_a + C_w = 1{,}000{,}000$. Then the concentrations are obtained as ppm units (g/m$^3$).

Consequently, the properties needed to estimate the partition of a new organic chemical to environment in level I equilibrium are the following:

$M =$ molecular weight
$T =$ ambient temperature (to be given) (K)
$P =$ vapor pressure at $T$ K (mmHg) (or, if vapor pressure is not known, then $P$ is calculated by

$$\log P = -52 \times (A/T) + B, (B = 9 \text{ for organic matter}),$$
$$A = 6.12 \times (273.1 + Bp)/52.23,$$

where Bp = boiling point °C
$S =$ solubility to water mg/l (ppm)
Ocf= fraction of organic matter in soil (to be given). In subarctic countries like Finland, Ocf is as high as 0.1, and in tropical areas as low as 0.02.

In level I calculation, no further input to or any outflow from the unit world for the material is assumed. The compound is assumed to be persistent in all compartments. Therefore, level I estimation is not useful for exposure assessment unless degradation properties of the chemical are not added by some method (see below).

Level II calculation[9] balances a constant input of material by a constant outflow. The outflow is the result of all possible reactions that occur in each compartment. The most important reactions are photodegradation (in air and water), hydrolysis (in water), and biodegradation (in soil and water). These all represent removal from the total unit world, while the calculation model is of equilibrium. The removal rate from a compartment ($i$) is:

$$R_i = \sum [M_i \Sigma k_i],$$

where $M_i$ = number of moles in compartment $i$ and $\sum k_i$ = summation of dissipating rate constants in compartment $i$, and the total removal rate $R_t$ from the unit world is the double summation:

$$R_t = \Sigma_i [M_i \Sigma k_i] \tag{3.2}$$

Residence time for a compound in the unit world is estimated from the total moles (Equation 3.1) and from total rate of removal (Equation 3.2).

The concentration in water at steady state is estimated by:

$$C_W = \frac{\text{Input (mol / day)}}{\Sigma_i (V_i D_i \Sigma K_i)}, \text{where}$$

$K_i$ = appropriate partition coefficient and
$D_i$ = appropriate density

When $C_w$ has been determined, the remaining concentrations may be estimated using appropriate partition coefficients.

Level III calculation[10] is a fugacity model which, in addition to steady state partitioning (level I) and reactions (level II), considers advection and intercompartmental transfers.

In level III procedure, partitioning is determined as Z values (fugacity capacities) for each compartment. Then, reactions (hydrolysis, oxidation, photolysis, and biolysis) are expressed as first order rate constants k [$h^{-1}$] and summed for each phase.

Advection terms are included for air (i = a) and water (i = w) flows in the form:

$$\text{Chemical inflow } Gi \times Cbi = Gi \times Zi \times fbi$$
$$\text{Chemical outflow } Gi \times Ci = Gi \times Zi \times fi$$

where Ga = air inflow and outflow rate of $5 \times 10^7$ $m^3/h$ corresponding to an air residence time of 5 days, Gw = water inflow and outflow rate of $2.9 \times 10^3/h$ the residence time being 100 days, Cbi = background inflow concentration (normally zero), and fbi = background fugacity (normally zero)

Two interphase transport processes are considered. Diffusive transport of solutes (driven by fugacity differences) are expressed by two transport coefficient values $D_{ij}$ and $D_{ji}$. The corresponding transfer rates are $D_{ij} \times f_i$ from *i* to *j* and $D_{ji} \times f_j$ from *j* to *i*. The net transfer rate *N* mol/h is the algebraic sum of the two counterdiffusion rates expressed as $N = (D_{ij} \times f_i) - (D_{ji} \times f_j)$.

The second transport process considered is the material transport. It is quantified as the product of the source volumetric transfer rate $G_i$ $m^3/s$ and the source concentration $C_i$ $mol/m^3$. The rate expression is given by:

$$N = G_i \times C_i = G_i \times Z_i \times f_i = D_{ij} \times f_i$$

In the six-compartment model, there are five diffusion (two directions each) and two mass transfer processes to be added:

Diffusion Processes: air $\rightleftharpoons$ water; air $\rightleftharpoons$ soil; water $\rightleftharpoons$ sediment; water $\rightleftharpoons$ suspended sediment; water $\rightleftharpoons$ biota

Mass Transfer Processes: soil $\rightarrow$ water; suspended sediment $\rightarrow$ sediment

The level III calculations for selected environmental chemicals give behavior profiles that are consistent with the reported fate of them in the environment.[10] For example, trichloroethylene concentrations in the Liverpool area in 1975 were

| Compartment | Model mg/m$^3$ | Observed mg/m$^3$ |
|---|---|---|
| Air | .0014 | 0.006 |
| Water | 0.094 | 6 (max) |
| Soil | 0.570 | — |
| Sediment | 0.450 | 0.1 |
| Fish | 0.880 | 0.1–10 |

### 3.4.3 Other Exposure-Assessment Procedures

A quick preliminary evaluation for possible environmental risk for a new chemical is often needed when the available data do not allow the use of a complete modeling process such as level III fugacity calculations. Therefore, other systems have been applied to get a first estimation of exposure.

In an approach suggested to the European Community (EC), the level II and III calculation is replaced by the use of point factors that represent persistence in each compartment.[11] Usage patterns and bioaccumulation were also used as points. The exposure factors were generated for three principal compartments:

| | |
|---|---|
| Factor for soil | $E_s = (C_s \times P_s)/U$ |
| Factor for water | $E_w = (C_w \times P_w \times \mathrm{BA})/U$ |
| Factor for air | $E_a = (C_a \times P_a)/U$ |

where

$U$ = the use pattern; 1 = open, 2 = semiopen... 5 = closed use
$P_s$ = biodegradation; 1 = fast, 2 = slow
$P_w$ = hydrolysis; 1 = fast, 2 = slow
$P_a$ = photolysis; 1 = fast, 2 = slow
BA = bioaccumulation factor; 2 = Bcf < 100, 4 = Bcf 100 to 1000, 8 = Bcf 1000 to 10,000, 16 = Bcf > 10,000

In a Swedish suggestion,[12] the assessment is totally based on point values. The exposure-related information that is taken into account in this procedure is

1. Uses (metric tons per year)
2. Amount emitted (%)
3. Primary recipient (air, soil, water, or purification plant)
4. Expected main target in environment (soil, water, soil + water)
5. Mobility and persistence
6. Bioaccumulation potential (log $K_{ow}$)
7. Bioavailability

Other procedures in the prediction of the environmental fate are consideration of structure-activity relationships, comparison to the environmental behavior of closely related compounds, and observed concentrations of the compound itself in the environment. These methods are described later in this book.

## 3.5 ESTIMATION OF THE EFFECTS

Structure-activity relationships (SARs) and comparison to the known related structures give the first idea of possible effects of a new chemical. For a compound released to the environment and having persistence and other high-exposure properties, however, toxicity testing is needed also.

Acute toxicities in certain test animals are the most frequently available effect-related data for chemicals. Test protocols for human toxicity are used with terrestrial animals, and results are applicable to the evaluation of ecotoxicity. However, effects upon other animals, microbes, and plants, including acute and long-term toxicity to individuals and effects upon populations, communities, and the ecosystem are also relevant to an adequate evaluation of environmental risks of a persistent compound. Complete studies to fulfill these requirements are too time-consuming and expensive to be practical.

To meet a reasonable minimal predictability, a set of standard tests, e.g., following the Organization for Economic Cooperation and Development (OECD) Guidelines,[7,8] are normally required by chemical control authorities. For preliminary risk estimation of, for example, a new pesticide, the following set of data from standard acute toxicity tests could be used:[13]

1. Mutagenicity test I (e.g., Ames)
2. Mutagenicity test II (e.g., Chromosomal Aberration)
3. Skin sensitization
4. Mammalian orl $LD_{50}$ (oral, rat)
5. Mammalian inh $LC_{50}$ (inhalation, rat)
6. Earthworm soil $LC_{50}$
7. Soil microbial $ED_{50}$
8. Birds orl $LD_{50}$
9. Bees orl $LC_{50}$
10. Fish 96h $LC_{50}$
11. *Daphnia* 48h $LC_{50}$
12. Algae growth inhibition $EC_{50}$

where

D = dose; C = concentration; L = lethal; E = effective.

If only ecotoxicity is evaluated, the data relevant to human toxicity (points 1 to 3) can be omitted. The control manager can give effect points for each test result and can summarize the points to effect factors (EF). Effects ($EF_i$) and exposure ($E_i$) factors can be calculated separately for each compartment (soil, water, and air). Then, risk factors for each compartment [(*X* (soil), *Y* (water), and *Z* (air)] are calculated:

$$X = EF_s \times E_s; Y = EF_w \times E_w; \; Z = EF_a \times E_a$$

and normalized to factors $S_x$, $S_y$, and $S_z$ (setting the maximal $X,Y$, and $Z$ values to 1.000 and the minimal values to 0.000).[11,13]

Preliminary conclusions about the indicated risk can be automatically reached from the factors and given, for example, as great ($S_i \geq 0.55$), potential ($0.55 > S_i \geq 0.30$), or no risk ($S_i < 0.30$). An example of a microcomputerized estimate output[13,14] is given in Figure 3.5.1.

If the screening methods described above indicate no environmental risk for a chemical, no special restrictions or bans for its use are executed. However, monitoring studies in environment might later expose an unexpected risk. Then the chemical must be taken back to the control measures. If a risk is indicated, more ecotoxicological research is needed and the uses of the chemical must be restricted.

In special cases, chemicals are regulated in more detail. For example, environmental assessment of pesticides to protect ground-water and soil life needs more studies and administrative measures than chemicals in general.[15]

## 3.6 RISK MANAGEMENT PROCEDURES IN SOME COUNTRIES

The general laws and their implication for environmental chemical control were first developed in OECD countries. The Toxic Substances Control Act went into effect in the U.S.A. in 1976; it is the basis for controls applied by the Environmental Protection Agency. Similar activities and control offices have been established in most industrialized countries. For example, EC directive 79/831, since September 1981, obliged all member countries to classify new chemicals with respect to their potential hazard to environment and human health. Nordic countries cooperated in efforts to find the most reliable methods for chemical control.

National efforts are supported by international cooperation in research and harmonization of environmental risk management. United Nations organizations (e.g., the United Nations Environmental Program [UNEP] and the World Health Organization), OECD, and EC are active in this field.

ECOTOXA. Evaluation of the environmental risk of LINDANE preliminarly for its acute toxic hazard to soil, water and air wildlife species. Label -1 = parameter was not known/used.

Given or calculated physical/chemical properties:

| Molecul. mass M | Water solubil. S ppm | Bcf ! | Temperat. t | Vap.pressure mmHg | Melting point | Boil. point |
|---|---|---|---|---|---|---|
| 290.9 | 10.0000 | 96 | 20.0 | 0.000465 | -1.0 | -1.0 |

! = Bcf was calculated from Kow = 1995.3

LINDANE IS DISTRIBUTED IN LEVEL I MODEL (NEELY-MACKAY) to the four comparments of the unit world air (a), water (w) soil (s) and biosphere (b; fish) as parts per million (Ni):

| Org.Cfr | t deg.C | Na | Nw | Ns | Nb | Henry const |
|---|---|---|---|---|---|---|
| 0.100 | 20.0 | 229600 | 361958 | 408407 | 34.67 | 0.00074026 |

Point values for Exposure calculation:

| Cs | Cw | Ca | BA | Bio | Hyd | Fot | Ps | Pw | Pa |
|---|---|---|---|---|---|---|---|---|---|
| 3 | 3 | 2 | 2 | 2 | 2 | 2 | 2 | 3 | 2 |

BA 2 (Bcf < 100), 4 (<1 000), 8 (<10 000) or 16 (>10 000)
Bio,Hyd,Fot,Ps,Pa = 1 fast, 2 slow degradation (Pw 1-3).

Exposure factors calculated from the values:

| | |
|---|---|
| Exposure from soil/sediment | Es = 6 |
| Exposure from water | Ew = 15 |
| Exposure from air | Ea = 4 |

Toxicity data used for determination of the Effects:

| Test and symbol of its result | | Dimension | | Value | Ef-points |
|---|---|---|---|---|---|
| Mammalian acute oral | LD50 | Mor | mg/kg | 76.000 | 2 |
| Mammalian acute inhalation | LC50 | Min | mg/l | -1.000 | 2 |
| Toxicity to earthworms | LC50 | Ewt | ug/cm2 | 0.700 | 3 |
| Toxicity to soil microbia | ED50 | Smt | mg/kg | 5.000 | 2 |
| Toxicity to birds | LD50 | Brt | mg/kg | 250.000 | 1 |
| Toxicity to bees | LD50 | Bet | ug/bee | 0.200 | 2 |
| Toxicity to fish | LC50 | Ftx | mg/l | 0.090 | 3 |
| Toxicity to Daphnia | LC50 | Dtx | mg/l | 0.700 | 2 |
| Algae growth inhibition threshold | | Agi | mg/l | 0.300 | 3 |

Effect factors calculated from the values above:
Ef's = 7 Ef'w = 8 Ef'a = 5

Risk factors X,Y,Z and their normalized values Si:

| Soil: X | Sx | Water: Y | Sy | Air: Z | Sz |
|---|---|---|---|---|---|
| 42.0 | 0.765 | 120.0 | 0.220 | 20.0 | 0.333 |

The following conclusions can be drawn:
LINDANE is a great risk to soil environment
LINDANE is a potential risk to air environment

**Figure 3.5.1** Sample output of ECOTOXA estimation.

## 3.7 SUMMARY

The environmental risk of a chemical is the probability of its harmful effect, which can be estimated as the product of exposure of the target species and toxic potential of the chemical (dose-response ratio). Exposure assessment is preliminarily done using fate modeling from the known properties of the chemical, and verified by analyses in the environment. Effects are assessed by laboratory tests and verified by environmental observations. Integrated procedures for assessment and management of the environmental risk of new chemicals have been developed in industrialized countries.

## 3.8 REFERENCES

1. **Saxena, J., Ed.** *Hazard Assessment of Chemicals,* Vol. 5 (Washington, DC:Hemisphere Publishing Co., 1987).
2. **Vouk, B., C. Butler, D.G. Hoel, and D.B. Peakall, Eds.** *Methods for Estimating Risk of Chemical Injury: Human and Non-Human Biota and Ecosystems* (New York:Wiley, 1985).
3. **Malmfors, T.** "The Use of Toxicological Data for Risk Assessment, Now and in the Year 2000." *FEST-Toxicology in Europe* (Cambridge:Elsevier, 1986), pp. 9-11.
4. "Risk Assessment in the Federal Government: Managing the Process," National Research Council U.S. (Washington, DC:National Academy of Science Press, 1983).
5. **Barnes, D.G.** "Characterization of the Risks Posed by CDDs and CDFs," *Chemosphere* 18:33-39 (1989).
6. **Richardsson, M.L., Ed.** *Risk Assessment of Chemicals in the Environment* (London:The Royal Society of Chemistry, 1988).
7. "Guidelines for Testing of Chemicals" (Paris:OECD, 1981), and Addendum Issues currently released.
8. "Data Interpretation Guides for Initial Assessment of Chemicals" (Paris:OECD, 1984).
9. **Neely, W.B., and D. Mackay.** "Evaluative Model for Estimating Environmental Fate," in, *Modeling the Fate of Chemicals in the Aquatic Environment,* K. L. Dickson et al., Eds. (Ann Arbor, MI:Ann Arbor Science, 1982), pp. 127-143.

10. **Mackay, D., S. Patterson, B. Cheung, and W.B. Neely.** "Evaluating the Environmental Behavior of Chemicals with a Level III Fugacity Model," *Chemosphere* 14:335-374 (1985).
11. **Schmidt-Bleek, F., W. Haberland, A.W. Klein, and S. Caroli.** "Steps Towards Environmental Hazard Assessment of New Chemicals," *Chemosphere* 11:383-415 (1982).
12. **Landner, L.** "Kemiska Ämnens Miljöfarlighet. ESTHER manual för inledande bedömning," Naturvårdsverket Rapport 3243, Solna (1987).
13. **Paasivirta, J.** "Baskets, ecorisk and ecotoxa," *Chemosphere* 17:1655-1662 (1988).
14. **Paasivirta, J.** "Ecotoxa: Microcomputerized Risk Estimation of Chemicals," in, Proceedings of the 1st European Conference on Ecotoxicology, Copenhagen, Denmark, October 17-19 (1988), 406-409.
15. **Klein, A.W., L. Fiedler, B. Muhs-Bonte, C. Schüter, W. Klein, and W.Kördel.** "Environmental Assessment of Pesticides — A Conceptual Approach," *Chemosphere* 17:1391-1410 (1988).

# 4

# Structure and Toxicity

## 4.1 INTRODUCTION

Ecotoxicologists must utilize pharmacological principles governing uptake of substances by different routes, accumulation, steady-state levels, acute and chronic damage by chemicals, and genetic, carcinogenic, and teratogenic effects.[1] However, extensive use of pharmacological/toxicological studies for all chemicals released due to usage or as waste emissions into the environment is impractical. But hazard assessment of a new organic chemical can utilize the structure-activity relationship (SAR) as a fast primary screening where most suspicious molecules are selected for closer investigations. SAR has developed into a valuable scientific tool in the research and development of drugs and pesticides where new structures with potentially useful properties have been sought. The development of chemical weapons and the harmful side effects of drugs and pesticides have motivated further studies of SAR. Also, the increasing use of chemicals in industry, agriculture, forestry, and the household has induced studies in which the knowledge about the structure-related toxic effects of chemicals can accumulate.[1-4]

According to Paracelsus (1493–1541), "All things are poisons, for there is nothing without poisonous qualities. It is only the dose which makes a thing a poison." It is true that the toxic effect of a chemical is dose related. However, the harmful dose level is also related to the toxic activity and ability of the toxic molecule to penetrate to the target organ.

## 4.2 GENERAL DEPENDENCE OF TOXICITY ON STRUCTURE

Local toxic effects on tissues depend on the physical and chemical properties of the chemical. High chemical reactivity is a typical structure-dependent property that can cause symptoms such as irritation, swelling, and burning of skin. Systemic toxicity can be caused by compounds that have suitable physical or chemical properties for a high penetration rate in living organisms.

Environmental exposure of the species is a very important factor in toxic effects. Small, spherically shaped nonpolar molecules are volatile and consequently can pose a high degree of exposure through inhalation. Their toxic risk is therefore higher than that of other, less volatile, compounds having similar $LD_{50}$ values.

Lipophilicity is the major factor in high exposure through the skin. In contrast, hydrophilic compounds can penetrate into animal or human body tissues from the alimentary tract. The maximal penetration occurs by small molecules that are lipophilic but also possess some structural polarity, e.g., are able to form hydrogen bonds with organically bound oxygen or nitrogen atoms.[1,4]

Lipophilicity-polarity-related biological activity can be modeled by the octanol-water participation coefficient ($K_{ow}$). The highest activities appear by certain optimum values of $K_{ow}$ in each series of related active substances.[5]

Biological persistency, including slow metabolism, can lead to accumulation of a compound to toxic level in certain tissues. In lipid-rich tissue, the most persistent compounds are some methyl-metals and hydrocarbons or their analogues like halogenhydrocarbons. In other tissues, the ability of the substance to form covalent bonds increases the persistency. In many cases a metabolite is more persistent than the original chemical. For example, DDE is a very persistent metabolite of the insecticide DDT (Figure 4.2.1). Both DDT and DDE are highly lipophilic and also slightly polar, which gives them a very high penetration potential with reference to animal tissues.

Modification of the basic structure causes regular changes to the biological activity. The following rules have been developed:

- Introduction or extension of an alkyl group increases lipophilicity, which often appears as increased uptake by absorption.

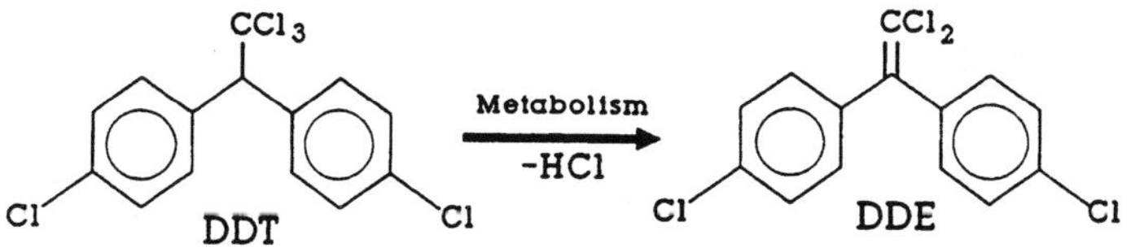


**Figure 4.2.1** Reaction of the insecticide DDT to DDE by chemical or metabolic loss of HCl.

- Branching of the alkyl chain makes the oxidative metabolism more difficult.
- Cycloalkyl groups increase the absorption rate by facilitating van der Waals bonding.
- Halogen atoms on the carbon skeleton increase lipophility. Often they also block sites of hydroxylative metabolism and therefore make the structure more persistent.
- Acylation or alkylation of –OH or –NH groups decreases the polarity. Consequently, absorption properties are changed and the molecule becomes more persistent.
- Metabolic methylation as a rule decreases toxicity of an organic compound but can produce more lipophilic and persistent substances.

The structure-specificity of a toxic effect varies considerably. Metabolic disturbances caused by acids, bases, and salts are very nonspecific by structure. Small molecular solvents have a nonspecific narcotic effect in the brain. In general, structurally nonspecific substances that are present in the same proportional saturation have the same thermodynamic activity and the same degree of biological activity.[1]

Penetration of an agent to the brain depends in a rather nonspecific manner on molecular properties. When the polarity is weak and the agent is relatively nonionizable, the compound is probably capable of breaking through the blood-brain barrier. Many biological effects depend on the structure of the active molecule, however. Structure-specific toxicants can be antimetabolites. These are molecules that resemble the important building blocks of biomolecules by structure enough to substitute for them in production of vitamins, coenzymes, or other essential moieties. Blocking of the bioaction follows as a toxic effect. Another common structure-specific influence takes place at the receptors.[1]

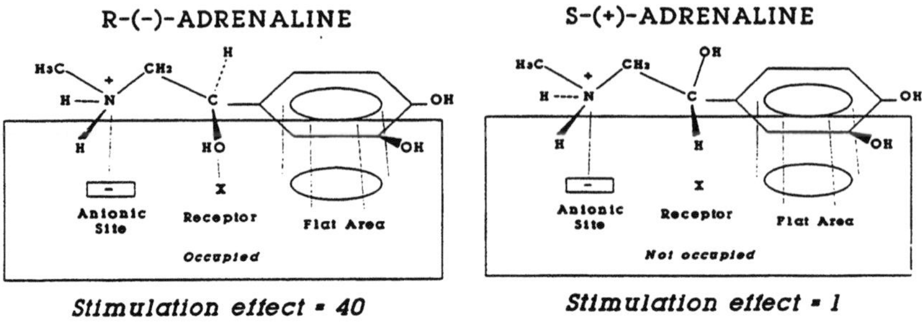


**Figure 4.4.1** Enantiomers of adrenaline (epinephrine) on an adrenergic receptor. A high stimulation effect needs attachment to all three receptor sites. Desoxyadrenaline, which lacks the alcoholic OH, has the same low stimulation effect as S-(+)-adrenaline.[6]

## 4.3 DEFINITION OF RECEPTOR

A receptor is an active structural unit (small area) of a cell membrane protein, enzyme, nucleic acid, or other biomacromolecule. The receptor is like a lock, and a small transmitter molecule of a life process is like a key that fits perfectly into that lock. If a molecule of a foreign (xenobiotic) chemical has a similar structure, size, and polarity, it may act as a "false key." Binding of a xenobiotic to the receptor can block a useful step of metabolism, execute a harmful metabolic process, or modify (accelerate or inhibit) nervous signal transmission. Thus the binding might cause acute or chronic poisoning or promote teratogenic or carcinogenic effects.

## 4.4 CHIRALITY AND RECEPTORS

Optical antipodes, as a rule, have different physiological activities. Receptor theory explains the rule quite satisfactorily. Adrenergic receptors in nervous cell membranes (Figure 4.4.1) consist of three structural units:

1. The anionic site (phosphate group), which binds an amino nitrogen by ion pair formation (after transfer of a proton from –P–OH to –NHR).
2. The active point (called the receptor), which attaches to the OH group.
3. A flat area, which attracts the benzene ring to a parallel orientation.

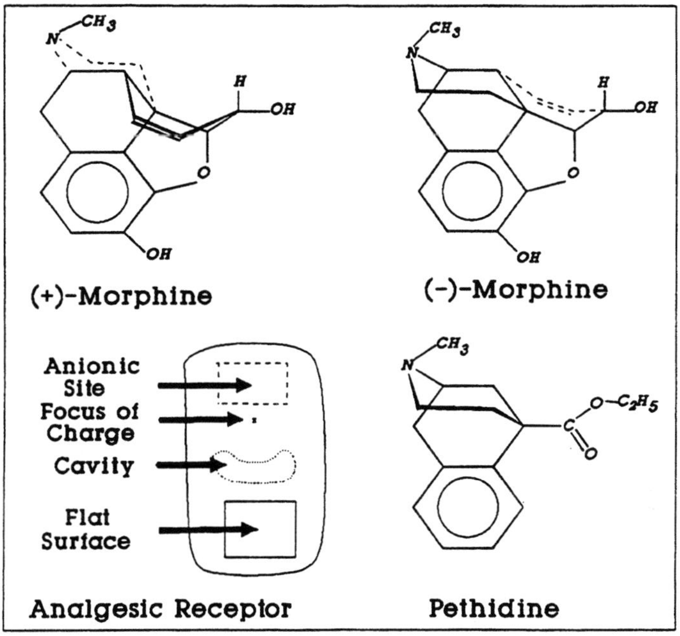


**Figure 4.4.2** Schematic picture of the analgesic receptor and spatial structures of (+)-morphine, (–)-morphine, and pethidine.

A very good example of a receptor-specific drug or toxicant is morphine. Only its naturally occurring (–)-enantiomer is physiologically active. (–)-Morphine and structurally similar compounds interact with the analgesic receptor in the brain (Figure 4.4.2).

Interaction of the analgesic drug molecule with a receptor involves attachment at three points. The anionic site forms an ion pair with the alkylamino group. Part of the piperidine ring fits into the cavity. Finally, the flat area of the receptor attracts the benzene ring. The other structural parts of morphine do not participate in the analgesic activity. This is proven by the high activity of a simple synthetic product pethidine (discovered by Eisleb in 1937). Mirror-image enantiomer (+)-morphine is totally inactive, because its molecule cannot conform to the proper spatial arrangement.

## 4.5 CHOLINERGIC RECEPTORS

### 4.5.1 Cholinergic Transmission of Nerve Signals

Adrenaline and related catecholamines (noradrenaline, do-

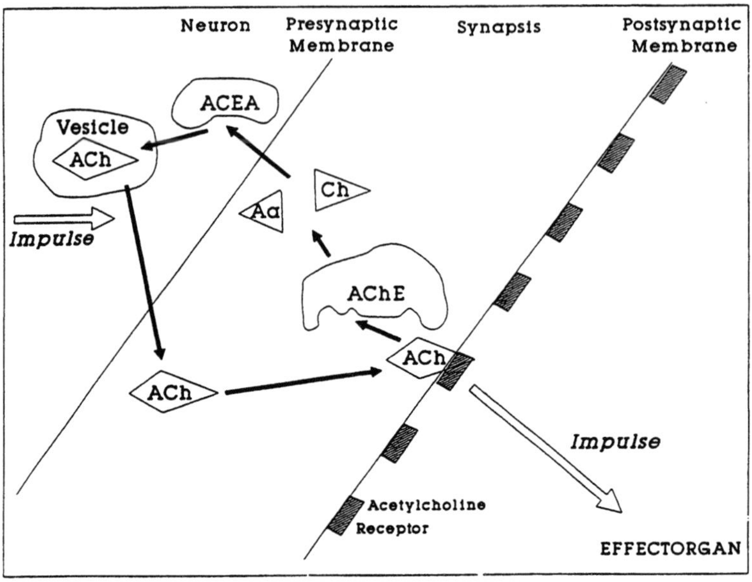


**Figure 4.5.1** Cholinergic nerve impulse transmission in synapsis (nerve endings or ganglia). For abbreviations, see text.

pamine) are examples of neurotransmitters, which mediate nervous impulses by receptor attachment (see above). Another type of neurotransmitter is acetylcholine (ACh). It mediates the transmission of impulses in the autonomic nervous system and at the neuromuscular junctions in the voluntary nervous system.

ACh is stored bound to protein in vesicles of a presynaptic neuron where a bioelectric current impulse liberates it to an active form. Next, ACh travels through the synapsis (the gap between nerve endings or ganglia) to irritate receptors on a postsynaptic membrane, and the nerve impulse proceeds (Figure 4.5.1).[6,7]

After transmitting the signal to the cholinergic receptor, ACh is rapidly hydrolyzed by action of the enzyme acetylcholine esterase (AChE) to yield inactive choline (Ch) and acetic acid (Aa). Another enzyme, acetyl coenzyme A (ACEA), joins Aa and Ch back to vesicle-bound ACh.

## 4.5.2 Direct Cholinergic Effect

Acetylcholine receptors are located in the postsynaptic membrane and in the enzymes AChE and ACEA. The mem-

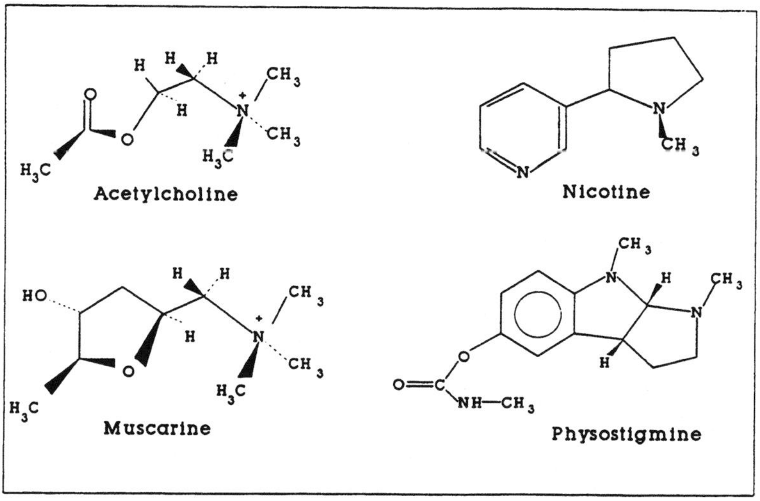


**Figure 4.5.2** Natural cholinergic compounds.

brane receptors receive irritation, which leads to continuation of the impulse towards the effect organ. Xenobiotics, which have similar structures, can bind to the receptor and cause irritation that is not interrupted by AChE and therefore gives toxic symptoms. Some structures that have such direct cholinergic effect are shown in Figure 4.5.2.

Toxic effects of the natural products nicotine, muscarine, and physostigmine rise from the structural similarity to acetylcholine (methylated nitrogen at 0.7 nm distance from oxygen or other polar group in specific spatial arrangement). This SAR was utilized in drugs for the stimulation of inner organs and muscles. Nicotine is an alkaloid that perhaps is produced by plants to ward off pests, and it is used by man as an insecticide. Consequently, effective pesticides were developed following SAR principles of the cholinergic structures. Also the carbamate group of physostigmine (lower left part of formula in Figure 4.5.2) gave rise to drugs and pesticides with –OCONHR substitution.

## 4.5.3 Indirect Cholinergic Effect

Physostigmine and most synthetic cholinergic agents in use as drugs or pesticides have an indirect influence to the signal

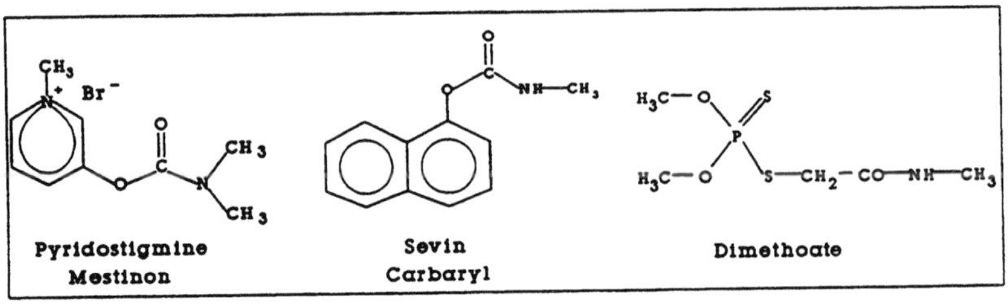


**Figure 4.5.3** Synthetic cholinergic agents: drug (mestinon) and two widely used insecticides (sevin and dimethoate).

transmission process of ACh. Three agents of that kind are shown in Figure 4.5.3.

The indirect effect includes binding of the agent to the receptor of the enzyme (AChE) and thus blocking of the hydrolysis of ACh. This cholinesterase inhibition causes the same symptoms as the direct irritation effect — the irritant now is the nondegradated ACh itself. Inhibition by carbamates is, as a rule, reversible: the receptor recovers in a relatively short time.[6]

The insecticide carbamate aldicarb is extremely toxic (its oral $LD_{50}$ is less than 1 mg/kg bodyweight). Even more toxic are some organophosphates that cause an irreversible cholinesterase inhibition. Some of these agents with effective exposure properties have been developed into "nerve agents" for chemical warfare.[2,3] Structures of aldicarb and two nerve agents are shown in Figure 4.5.4.

## 4.5.4 Anticholinergic Agents

The same SAR is effective as an antidote to cholinergic poisonings. Some natural and synthetic molecules have the structural properties to bind the acetylcholine receptor on the postsynaptic membrane, but they are not irritating to the receptor. Therefore, binding leads to blocking the signal transmission and prevention of the poisoning by cholinergic toxins. While the anticholinergic agents are antidote drugs against many poisons, by themselves they are toxic, since they inhibit the normal signal activity. An example of a natural and a synthetic anticholinergic agent is given in Figure 4.5.5. The alkaloid atropine is excreted by the flower *Atropa belladonna*. The extract was used in ancient Egypt as eye drops that dilated the

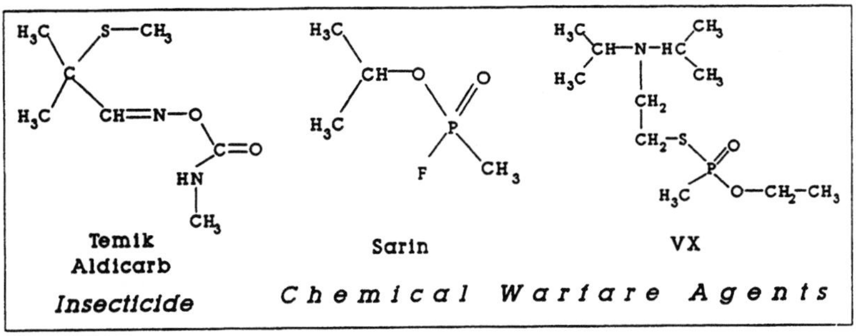


**Figure 4.5.4** Structures of three supertoxic cholinergic agents. Supertoxic means an agent with $LD_{50}$ 1 mg/kg or less.

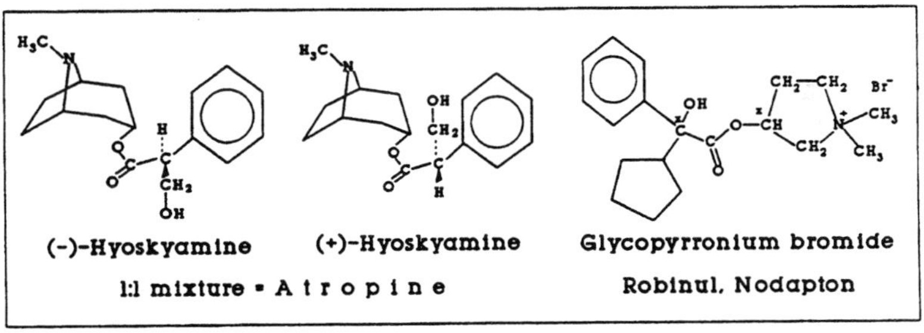


**Figure 4.5.5** Anticholinergic structures.

pupil of the eye (a symptom of receptor paralysis) and made the women more beautiful. Atropine injection can save a person's life after as much as an otherwise tenfold lethal exposure to a nerve agent or insecticide.

## 4.5.5 Reactions on the Receptor

To describe the irreversible, indirect cholinergic effect, the hydrolysis scheme of ACh on the receptor of AChE is first shown (Figure 4.5.6). The reaction of ACh involves a reversible acetylation of the phenolic OH of the tyrosine unit in the receptor. A nerve-agent molecule phosphorylates the OH group. The P–O bond formed is resistant to hydrolysis, and consequently the receptor is irreversibly blocked.

Receptors can be reactivated with special drugs such as pralidoxime (PAM; Figure 4.5 7).

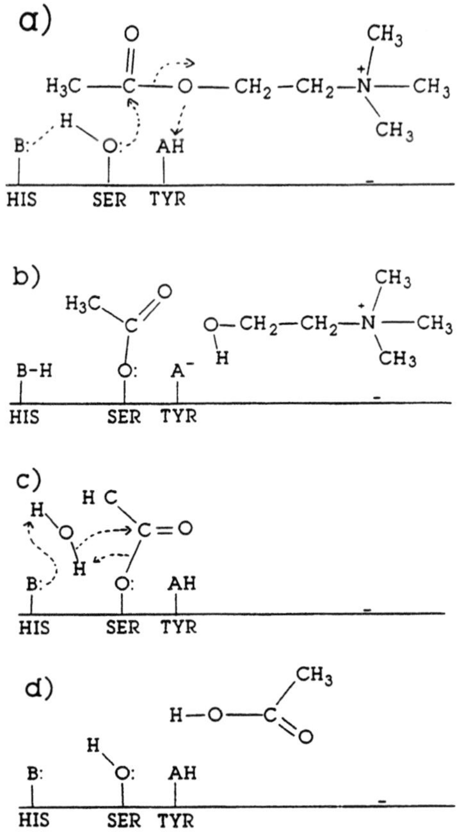


**Figure 4.5.6** Hydrolysis of acetylcholine (ACh) on the receptor of the acetylcholine esterase (AChE): (a) Imidazole ring of histidine activates the oxygen atom of serine, which, together with the phenolic OH proton of the tyrosine unit, conducts a two-point attack on the ACh molecule. (b) Oxygen atom of serine is acetylated and choline is released. (c) Water molecule is activated. (d) Acetic acid has been released by hydrolysis and the enzyme AChE is active again.

## 4.6 DIOXIN RECEPTOR

Hydrophobic xenobiotics, once absorbed by the organism, can be excreted only after metabolic conversion to polar products. In many cases the initial metabolic step is oxygenation on microsomal monoxygenase enzymes. For polycyclic aromatic hydrocarbons (PAHs), such an enzyme is cytochrome P-450 (and P-448), which occurs in many forms.[4,8] Increased exposure of liver cells to PAHs causes increased production and activity

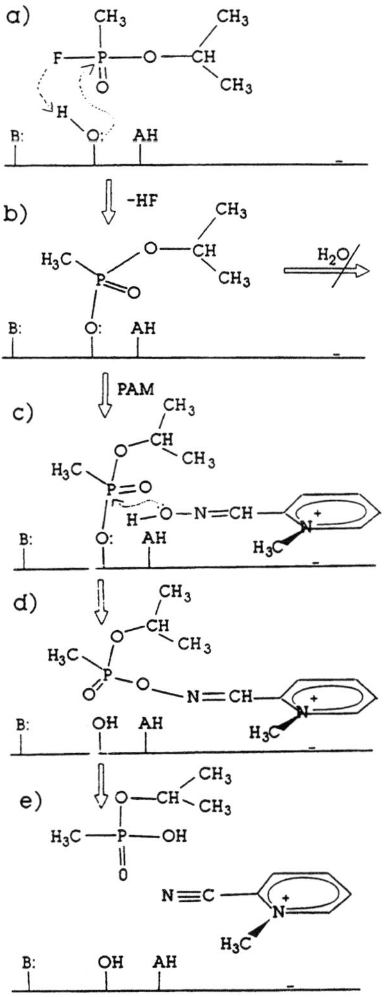


**Figure 4.5.7** Irreversible inhibition of the acetylcholine esterase receptor by a nerve agent and therapy with an oxime reactivator. (a) Sarine molecule on the receptor of AChE. (b) Receptor is poisoned by formation of the stabile phosphate, which is not hydrolyzed. (c) Added reactivator pralidoxime (PAM) activates the loss of phosphate. (d) An oxime phosphate is formed and the enzyme receptor released. (e) The oxime ester is degraded to nitrile and to a nontoxic phosphonate ester acid.

of these enzymes, e.g., aryl hydrocarbon hydrogenase (AHH) induction. The PAH molecule is bound to the planar receptors of the enzymes and oxidated there.

Other structurally similar xenobiotics can also attach to these receptors. The notorious "Seveso dioxin," 2,3,7,8-tetrachlo-

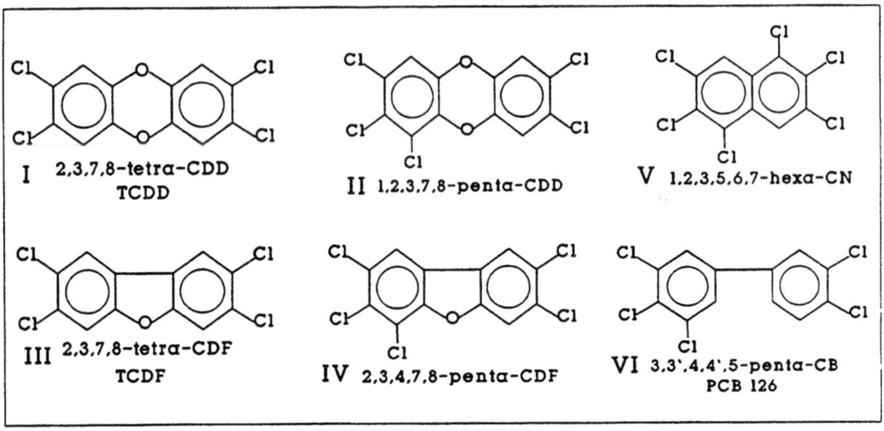


**Figure 4.6.1** Six common dioxin-type environmental toxicants.

rodibenzo-p-dioxin (TCDD), has a planar aromatic structure that obviously fits closely to the AHH receptors. Binding of TCDD or other structurally related dioxin-type toxicants causes induction of enzymes (e.g., AHH or ethoxyresorufin-O-dethylase [EROD]), thymic athropy, hypercreatosis, and a severe skin disease named chloracne in man. Very small doses of dioxins are fatal to many animal species.[9-13]

All aromatic compounds that can fill a planar rectangular space of 0.68 × 1.4 nm (a flat box 0.4 nm high) and have suitable surface polarity (four chlorines or three bromines in lateral positions are most active) belong to this class of dioxin-type poisons. The persistent, bioaccumulating molecules in this group are the most potent environmental toxicants known thus far. Examples of these xenobiotics are shown in Figure 4.6.1.

TCDD (I in Figure 4.6.1) is the most toxic manmade chemical known. Its oral $LD_{50}$ (acute toxicity) to guinea pig is 0.6 µg/kg bodyweight, and to rat and human about 30 µg/kg bodyweight.[13] The chronic toxicity is also very high. Dioxin II and the furans III and IV are only slightly less toxic when all effects are considered. Compound V is the most persistent chloronaphthalene in biota and, as laterally chlorosubstituted, might be of environmental concern. Compound VI is the most toxic polychlorinated biphenyl (PCB) congener.

Molecular graphics (description of molecules as their van der Waals surfaces) shows more closely the exact size and properties of the dioxin receptor (Figure 4.6.2).

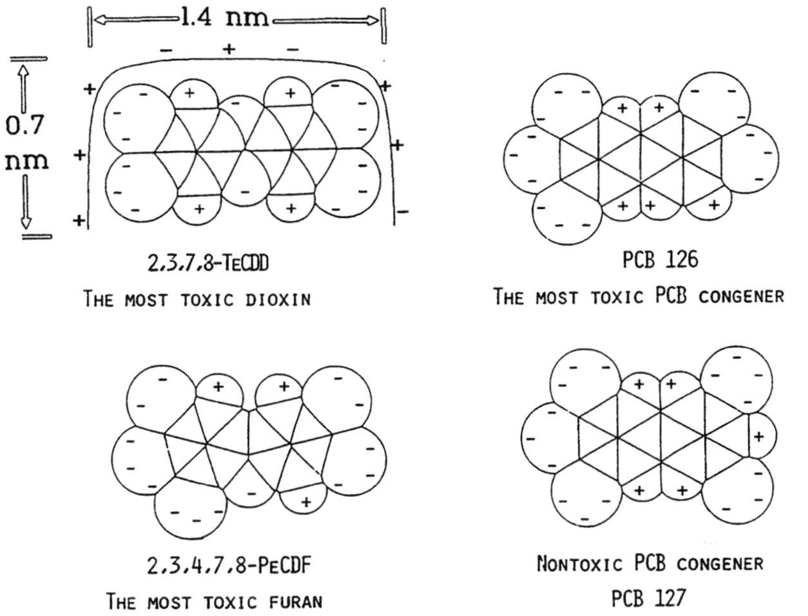


**Figure 4.6.2** Dioxin receptor and agent structures described as van der Waals surfaces and as electron-attracting (+) or repulsing (–) areas.

In addition to planar aromatics, those PCBs that can turn to a planar arrangement fit to the dioxin receptors. Their toxicity, however, depends very much on the locations of the chlorine substituent. When only one chlorine atom (4′) in the skeleton (of PCB 126) is changed (to 5′), AHH induction (for PCB 127) decreases to 1/50,000 (of that of PCB 126).[14]

Other dioxin-type toxic structures are, e.g., polyhalogenated anthracenes, biphenylenes, and azobenzenes.[13] Also, nonhalogenated planar structures such as indolecarbazoles and benzoflavones have the same structure-related toxic function.[12]

Molecular geometry and structure determinations of model compounds by spectroscopy and X-ray diffraction can be used to predict the toxic potential of an environmental chemical comparing the results to known receptor properties and SARs.[15]

A faster, useful screening method for possible dioxin-type toxicity a new planar aromatic chemical is the measurement of AHH or EROD induction. The results correlate fairly well with all other toxic effects and therefore give suitable reference values of toxic potency.[16,17] Relative potency of induction (toxic equivalency factor, TEF) for AHH in rat hepatoma E-4-IIE cell

**Table 4.6.1 Toxic equivalency factors (TEFs) for dioxins and furans**

| PCDDs and PCDFs | TEF/AHH[16,17] | TEF/NORDIC[18,a] |
|---|---|---|
| 2,3,7,8-TeCDD TCDD | 1.0 | 1.00 |
| 1,2,3,7,8-PeCDD | 0.0066 | 0.50 |
| 1,2,3,4,7,8-HxCDD | 0.034 | 0.10 |
| 2,3,7,8-TeCDF | 0.18 | 0.10 |
| 1,2,3,7,8-PeCDF | 0.029 | 0.01 |
| 2,3,4,7,8-PeCDF | 0.28 | 0.50 |
| 1,2,3,4,7,8-HxCDF | 0.20 | 0.10 |
| 1,2,3,6,7,8-HxCDF | 0.048 | 0.10 |
| 2,3,4,6,7,8-HxCDF | 0.10 | 0.10 |

[a] NORDIC TEF values are based on acute and long-term toxicity data for animals in evaluation by an expert group.[18]

lines by polychlorodibenzo-p-dioxins (PCDDs), polychlorodibenzofurans (PCDFs), and non- and mono-ortho-Cl coplanar PCBs is calculated:

$$\text{TEF} = \text{Relative potency of induction} = \frac{EC_{50}(\text{TCDD})}{EC_{50}(\text{Substance})}$$

Some comparisons of TEFs are listed in Tables 4.6.1 and 4.6.2.

To compare the environmental toxic significance of those xenobiotics whose TEFs have been determined, their analysis results in biota can be transferred to the toxic load of each chemical as toxic equivalents (TEQs) of TCDD:[16,19]

$$\text{TEQ} = \text{TEF} \times \text{concentration.}$$

In Finnish, Baltic, and Arctic environments, the coplanar PCBs, mainly PCBs 126 and 105, represent a vast majority of the known TEQ load of animals. The portion of PCDDs and PCDFs is significantly smaller except in seals, where it seems to be nearly as large as that of coplanar PCBs.[19-23]

## 4.7 SUMMARY

The ecotoxicity of a new organic chemical can be predicted from its molecular structure utilizing structure-activity rela-

**TABLE 4.6.2 Toxic equivalency factors (TEFs) for coplanar PCBs**

| Coplanar PCBs | PCB Nr | TEF/AHH[14,16,17] | TEF/Thym.tox[17] | a |
|---|---|---|---|---|
| 3,3′,4,4′-TeCB | 77 | 0.0021 | | 0.01 |
| 3,3′,4,4′,5-PeCB | 126 | 0.30 | 0.95 | 0.1 |
| 3,3′,4,5,5′-PeCB | 127 | 0.000007 | | |
| 3,3′,4,4′,5,5′-HxCB | 169 | 0.0012 | 0.010 | 0.05 |
| 2,3,3′,4,4′-PeCB | 105 | 0.00082 | 0.000087 | 0.001 |
| 2,3′,4,4′,5-PeCB | 118 | 0.000006 | 0.000058 | 0.001 |
| 2,3,3′,4,4′,5-HxCB | 156 | 0.000034 | 0.00050 | 0.001 |

[a] Recommendation of S. Safe in DIOXIN'90, Bayreuth, September 1990.

tionships (SARs). Modification of the known structure causes predictable changes to its exposure-related properties and toxic effects. Some effects are nonspecific by structure, but the most important known environmental xenobiotics exert their influence via metabolic receptors. Examples of such toxicants are cholinergic agents and dioxin-type planar aromatic compounds.

## 4.8 REFERENCES

1. **Goldstein, A., L. Aronow, and S. M. Kalman.** *Principles of Drug Action. The Basis of Pharmacology, 2nd ed.* (New York:Wiley, 1974).
2. SIPRI. *The Rise of CB Weapons* (Uppsala:Almqvist and Wicsell, 1971).
3. **Stöhr, R., et al.** *Chemishe Kampfstoffe und Schutz vor Chemishen Kampfstoffen* (Leipzig: Militärverlag der DDR, 1977).
4. **McKinney, J. D., Ed.** *Environmental Health Chemistry* (Ann Arbor, MI:Ann Arbor Science Publishers, 1981).
5. **Matsumura, F.** *Toxicology of Insecticides* (London:Plenum Press, 1985).
6. **Wilson, C. O., O. Gisvold, and R. F. Doerge.** *Textbook of Organic Medicinal and Pharmaceutical Chemistry* (Philadelphia:J.B.Lippincott Co., 1977).
7. **Ebel, S.** *Synthetische Arzneimittel* (Weinheim:Verlag Chemie, 1979).
8. **Cattabeni, F., et al., Eds.** *Dioxin. Toxicological and Chemical Aspects* (New York:Spectrum Publications Inc., 1978).

9. **Poland, A., E. Clover, and A. S. Kende.** "Stereospecific, High Affinity Binding of 2,3,7,8-Tetrachlorodibenzo-p-dioxin by Hepatic Cytosol; Evidence that the Binding Species is Receptor for Induction of Aryl Hydrocarbon Hydroxylase," *J. Biol. Chem.* 251:4936-4946 (1976).
10. **McKinney, J., and E. McConnell.** "Structural Specificity and Dioxin Receptor," in, *Chlorinated Dioxins and Related Compounds,* O. Hutzinger et al., Ed. (Oxford: Pergamon Press, 1982), pp. 367-381.
11. **Safe, S., L. W. Robertson, and L. Safe.** "Halogenated Biphenyls; Molecular Toxicology," *Can. J. Physiol. Pharmacol.* 60:1057-1064 (1982).
12. **Gillner, M., B. Fernström, J-Å. Gustafsson, J. Bergman, and C. Cambillau.** "A New Structure-Affinity Relationship for TCDD Receptor Binding," *Chemosphere* 15:1673-1680, (1986).
13. **Hutzinger, O., et al., Ed.** *Chlorinated Dioxins and Related Compounds* (Oxford:Pergamon Press, 1982).
14. **Mannila, E., and S. Kärelampi.** "Synthesis, Structure Proof and First Toxicity Test of Coplanar 3,3′,4,5,5′-pentachlorobiphenyl," SETAC, Transboundary Pollution, Tenth Annual Meeting, Toronto, October 28 - November 2 (1989), 275, and paper to be published.
15. **Grainger, J., V. V. Reddy, and D. G. Patterson, Jr.** "Molecular Geometry/Toxicity Correlations for Laterally Tetrachlorinated Dibenzo-p-dioxins by Fourier Transform Infrared Spectroscopy," *Chemosphere* 18:981-988 (1989).
16. **Kannan, N., S. Tanabe, and R. Tatsukawa.** "Toxic Potential of Non-ortho and Mono-ortho Coplanar PCBs in Commercial PCB Preparations: 2,3,7,8-T4 CDD Toxicity Equivalence Factors Approach," *Bull. Environ. Contam. Toxicol.* 41:267-276 (1988).
17. **Safe, S.** "Determination of 2,3,7,8-TCDD Toxic Equivalent Factors (TEFs): Support for the Use of the IN VITRO AHH Induction Assay," *Chemosphere* 16:791-802 (1987).
18. Nordic Expert Group. "Nordisk Dioxinrisk Bedömning," NORD 1988:49 (Kobenhavn:Nordisk Ministerråd, 1988).
19. **Paasivirta, J., J. Koistinen, E. Mannila, J. Tarhanen, P. J. Vuorinen, N. Kannan, N. Yamashita, S. Tanabe, and R.Tatsukawa.** "Coplanar PCBs in Finnish Wildlife,"

SETAC, Transboundary Pollution, Tenth Annual Meeting, Toronto, October 28–November 2 (1989), platform presentation 271, and a manuscript for Chapter 7 in *Coplanar PCBs* (Chelsea, MI:Lewis, to be published).

20. **Tarhanen, J., J. Koistinen, J. Paasivirta, P. J. Vuorinen, J. Koivusaari, I. Nuuja, N. Kannan, and R. Tatsukawa.** "Toxic Significance of Planar Aromatic Compounds in Baltic Ecosystem — New Studies on Extremely Toxic Coplanar PCBs," *Chemosphere* 18:1067-1077 (1989).
21. **Paasivirta, J., K. Mäntykoski, J. Koistinen, T. Kuokkanen, E. Mannila, and K. Rissanen.** "Structure Analyses of Planar Polychloroaromatic Compounds in Environment," *Chemosphere* 19:149-154 (1989).
22. **Koistinen, J., J. Paasivirta, and P. J. Vuorinen.** "Dioxins and Other Planar Polychloroaromatic Compounds in Baltic, Finnish and Arctic Fish Samples," *Chemosphere* 19:527-530 (1989).
23. **Koistinen, J.** "Residues of Planar Polychloroaromatic Compounds in Baltic Fish and Seal," *Chemosphere* 20:1043-1048 (1990).

# 5

# Sampling and Analysis of Trace Chemicals in the Environment

## 5.1 INTRODUCTION

Trace chemicals in the environment are studied to protect nature and human health. Of particular interest are xenobiotics, which, in bioaccumulating, can create significant hazard. Sampling and analyses of environmental chemicals are done, e.g., by the following activities:

1. Basic research in order to develop new sampling and analysis methods for environmental chemical traces.
2. Screening or survey of the occurrence, transport, and effects of persistent anthropogenic chemicals.
3. Monitoring of trends of known environmental toxicants using established standard methods.
4. Control of the emissions for regulatory purposes and for prevention of harmful waste spills.
5. Chemical hazard situations, accidents, etc., which need sampling and analyses of known and unknown hazardous chemicals.

Techniques include direct measurements from air, water, and surfaces, sampling of air, water, soil, and biota, handling of the samples, analytical cleanup procedures, determinations primarily by instrumentation, data reduction, and interpretation of the results using integrated environmental, sample, and concentration data.[1-3]

## 5.2 SAMPLING METHODS

### 5.2.1 General Aspects

Sampling with minimal and clearly defined error is essential for producing reliable results of environmental traces. Well-planned tools, pure devices, and careful procedures applied by well-trained personnel are needed to get representative and noncontaminated samples. The sample size must be large enough to obtain the necessary accuracy of determinations and provide for possible repetition of the analyses. As a minimal rule, two to three parallel samples should be taken from each point.

### 5.2.2 Planning and Carrying Out the Sampling

Sampling is included in the overall investigative plan where study methods, measured variables (endpoints), sampling places, and sampling frequency are established. The sampling plan should be based on an overall study program developed with a competent statistician.

Equipment for sampling is also included in the formal plan. Especially in cases of potential hazard to personnel samplers must wear rubber boots and gas masks and have all skin surfaces completely protected by rubber clothing. The sampling personnel must have explicit, detailed instructions on the sampling protocol and locations to be sampled.

It is essential to avoid contamination during the sample collection. The sample containers must be clean and of the correct type for the purpose. Plastic is an unsuitable bottle material for hydrophobic contaminants in water. Glass must be used instead. Air contaminants at the sampling site, such as dust and soot, might cause errors in sampling for the background levels of traces in air, water, soil, and biota. Fuel and exhaust gases of vehicles might cause errors in samples taken for metal or organic pollutant determinations. Plankton nets can collect impurities from boat paints, fuel, and oil if they are towed at the surface behind the boat. Fallout samples collected below an electric power line may collect copper and other metals from the wire, wire coatings, or support structures.

For some analyses, immediate pretreatment of the sample is needed. For example, one could arrest any potential biodegradation in a wastewater sample by the addition of sulfuric acid.

Before transportation to the laboratory, all samples must be labeled clearly. The number or name of the sample, date, location, sampling person, contact person, and address of the laboratory should be marked on the label and on the documents following the sample in transport. The purpose of the sample, e.g., which analyses will be done, should also be indicated on the label. Conditions and results of measurements done at the sample location must be marked on sample containers or on associated documents. To assess the extent of cross-contamination, zero-sample containers, adsorption tubes, etc., should be kept and transferred together with the actual samples for blank analyses.

In case of hazardous samples, the containers must be wiped to clean the outsides before packing them in larger wooden or metallic boxes filled with soft, chemically inert adsorbent material like activated charcoal. In any case, safety of the transportation must be considered.

### 5.2.3 Determination of Sample Size

To obtain a sufficient amount of data for reliable statistical analysis, the number of the samples from each place (or population) studied (sample size) must be determined in advance. The determination can be based on the relationships between standard error, sample size, population distribution, analysis method, and mean concentration of the trace substance determined. However, the exact relationships often depend on unknown factors such as the population variance. Therefore, some approximate guidelines must be used in most cases.[3]

In studies of emissions and in monitoring studies, as a rule, the coefficient of variation for single samples (cv) at different means of concentrations ($x$) are known. Then, for a given statistical accuracy of the results ($D$) the sample size needed ($N$) is

$$N = (\mathrm{cv} / D)^2 \qquad (5.2.1)$$

*Example:* Hexachlorobenzene contamination of two lakes is compared by analyzing populations of pike. An accuracy (deviation in the result as its fraction) $D = 0.3$ is known to be enough for determination of the statistically significant difference. The predicted mean concentrations ($x$) are about 10 ng/g. Previous studies have given cv = 1.4. From equation (5.2.1), one calculates the needed sample size $N = 22$.

### 5.2.4 Optimization: Composited Samples

The required large sample sizes to carry out statistically valid survey and monitoring studies of trace chemicals can make such programs prohibitively expensive. For a single sample, the analysis costs are often very much higher than the sampling costs. In those cases the costs can be reduced without losing significance in the results by using composite samples.[4]

The theory for the optimal sample composing is simple. The number of analyses (*a*) needed to obtain the same accuracy as *k* analyses of individual samples is

$$a = k \times \left[\text{lo} + (1 - \text{lo}) / n\right] \qquad (5.2.2)$$

where *n* is the number of individuals homogenized together to make a composited sample and lo is the fraction of laboratory variance (LV) in the total variance. In case of biological samples, total variance (*V* = square of the standard deviation *d*) consists of the laboratory variance (LV) and of the biological variance (BV):

$$\text{lo} = \text{LV} / V \qquad (5.2.3)$$

$$V = \text{LV} + \text{BV} = d^2. \qquad (5.2.4)$$

While the sampling and analysis costs are known, one can minimize the costs of a monitoring (or surveying) study using equation (5.2.2) if the laboratory variance (LV or lo) is known or can be predicted with reasonable accuracy. Prediction of LV or lo is possible only from the experiences of actual analyses. When large populations are studied, duplicate analyses from part of the samples can be done for the determination of LV:[4-6]

$$\text{LV} = \sum_{1}^{N} (\text{ds})^2 / (2 \times N) \qquad (5.2.5)$$

where ds is the difference between the concentration measured in the determination and in its duplicate. *N* is the number of duplicates.

LV has been observed to be very different in different laboratories (Figure 5.2.1). As a rough estimate, it can be seen to increase exponentially with decreasing mean of concentration $(x)$.[5,6]

The validity of equation (5.2.2) was experimentally proven by analyzing a large population (from a single catch) of herring as 25 individuals and as 12 composites of 3, 9 composites of 5, and 8 composites of 8 individuals. A duplicate determination of every fifth sample (composite) was done to obtain lo. Variances from the composite analyses were equal to or better than those calculated by equation (5.2.2) in 157 cases of the total 159. The economy of a monitoring study for chlorohydrocarbons in herring could be optimized by the following deduction:[4]

- Number of individual analyses to obtain the sufficient accuracy for one population was $k = 25$.
- Largest laboratory variance observed (for chlordanes) was taken to be lo = 0.286.
- Combinations of $n$ and $a$ that gave the same accuracy were calculated and compared to the cost, which was 15:– for each herring sampled and 450:– for each determination:

| $n$ | $a$ | Sample Costs | Analysis Costs | Total |
|---|---|---|---|---|
| 1 | 25 | 375:– | 11,250:– | 11,625:– |
| 2 | 16 | 480:– | 7,200:– | 7,680:– |
| 3 | 13 | 585:– | 5,850:– | 6,435:– |
| 4 | 12 | 720:– | 5,400:– | 6,120:– |
| 5 | 11 | 825:– | 4,950:– | 5,775:– |
| 6 | 10 | 900:– | 4,500:– | 5,400:– |
| 7 | 10 | 1,050:– | 4,500:– | 5,550:– |

Thus, the optimized procedure includes sampling 60 herring and analyzing them as 10 homogenates of 6 individuals. The analysis number $N = 10$ is large enough for most statistical comparisons with other populations. Saving of the costs related to 25 individual specimen analyses is $100 \times (11{,}625–5400)/11{,}625 = 53.5\%$.

In certain situations, using composite samples improves the limit of determination of a trace chemical due to the greater tissue mass available for each analysis.[7]

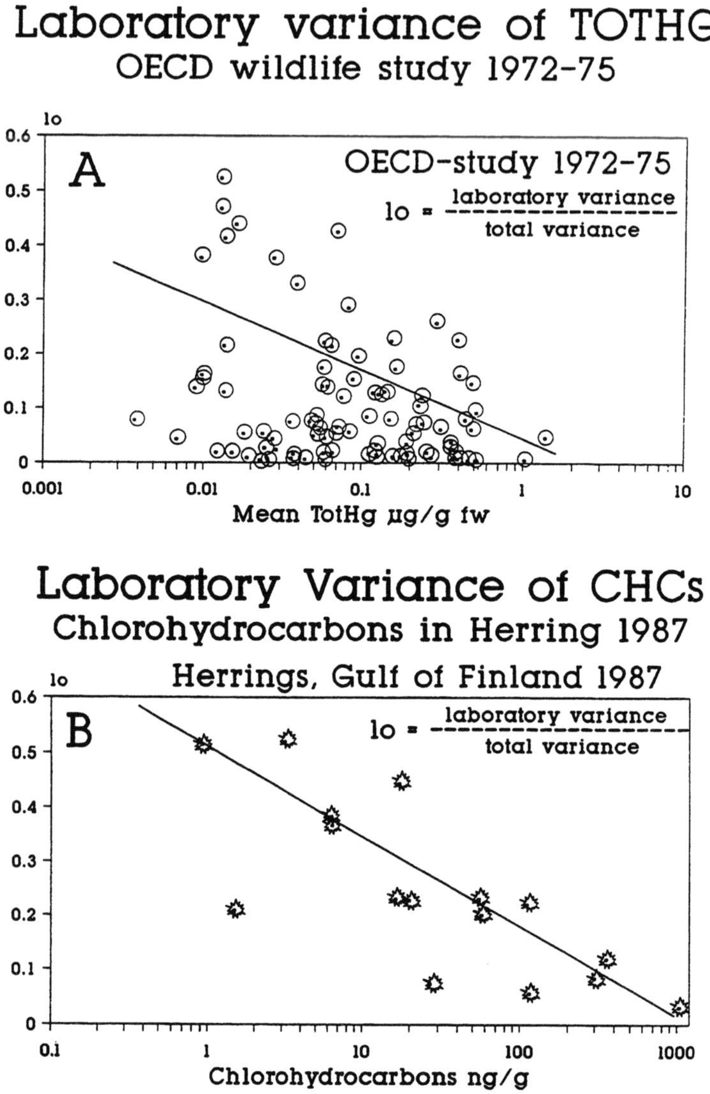


**Figure 5.2.1** Dependences of the laboratory variation (lo) on mean ($x$). (A) Mercury concentration in wildlife samples in some OECD (Organization for Economic Cooperation and Development) countries from 1972 to 1975.[5] (B) Chlorohydrocarbons in a single cast of herrings. Results in both fresh and lipid weight basis are included.[4] The scale is linear for lo but logarithmic for $x$.

### 5.2.5 Air Samples

Samples of solid particles, aerosols, and gases are taken from air. For trace analyses, the most frequently used sampling

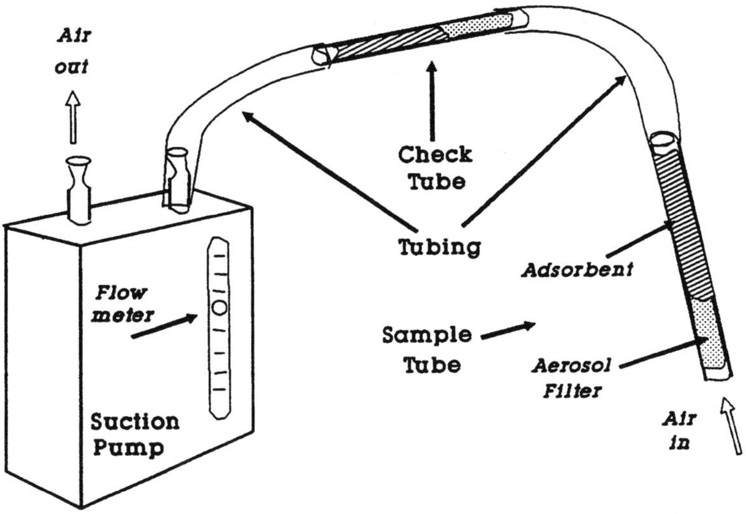


**Figure 5.2.2** Low-volume sampling of air to adsorption tubes.

devices are constructed for low-volume sampling over short time periods from point locations and for high-volume sampling over long time periods from larger volumes (e.g., from an airplane). The devices used for indoor air sampling in occupational health control (sacks, diffusion tubes, etc.) are useful outdoors only in cases of high concentrations (chemical accidents and other hazard situations).

Small-molecular gases in air are measured directly. For example, sulfur dioxide and hydrogen sulfide are measured by ultraviolet fluorescence, carbon monoxide by infrared spectrometer, and nitrogen oxides with a chemiluminescence device. Many toxic organic gases can be measured with electrochemical cells, flame-emission detectors, and ion cluster, piezoelectric, thermoacoustic, and photoacoustic sensors.

A general procedure for air sampling includes the use of a suction pump and filters. Some gases are collected also to chemically reactive (impregner) solutions in washbottles. For organic traces, however, filter tubes are handy in low-volume sampling (Figure 5.2.2).

The sampling pump must include a flow meter. Then the volume of the air sample (e.g., in ml) can be calculated from flow (ml/min) and sampling time (min).

Dust and aerosols are collected with membrane or glass fiber filters. Pore size for solid particles is preferably 0.8 μm, but for collection of aerosols, a smaller size is needed. An aerosol filter can be placed in the same tube before the adsorbent, which collects gaseous substances as shown in Figure 5.2.2.

The choice of adsorbent material depends on the desorption method used. In sampling for solvent desorption, activated carbon and polymeric resins (XADs) resins are widely used. More suitable materials for thermal desorption are the porous polymers such as Tenax GC (diphenyl-p-phenylene oxide), divinyl benzene polymers (Chromosorb 101 to 107), and polyvinyl pyrrolidones (Porapak N, P, Q, R, or S). Some molecular sieves are indifferent enough for chemically sensitive organic sulfides and hydrogen sulfides. For reliable determination, the adsorption tubes must be stored at low temperature and desorption/determination done as soon as possible.

If there is any doubt regarding the capacity of the sample tube, a check tube is used at sampling (Figure 5.2.2) after the actual analysis tube and is analyzed first. If the result is comparable to that from the blank sample determination, then all material to be analyzed in air has been collected in the sample tube and can be analyzed quantitatively.

Tenax tubes for thermal desorption need to be precleaned after filling by heating in pure nitrogen gas at 200°C for 2 hours or more. The sample tubes must be sealed gas-tight and stored frozen before the sampling. Some blank tubes must be transported together with the sample tubes to the sampling location and back to the laboratory for check determinations.

Air contaminants can also be studied from biological samples collected from the study area. Mushrooms, lichens on trees, and softwood needles accumulate some heavy metals and persistent organic pollutants from air.

Use of moss-bags as deposition gauges is a suitable method for air pollutant surveys, especially for metals. For example, Sphagnum moss (*S. girgensohnii*) can be collected from noncontaminated marsh and purified from extra trash and dead portions. After washing once with 0.5 M HCl and four times with deionized water, the moss is pressed and then stored to have a constant humidity and then weighed to make balls corresponding to 1 g of dry moss each in nylon nets. The balls are kept at the sampling area at a 2-m height for some weeks and then collected for analyses.

In chemical accidents and chemical warfare-like situations,

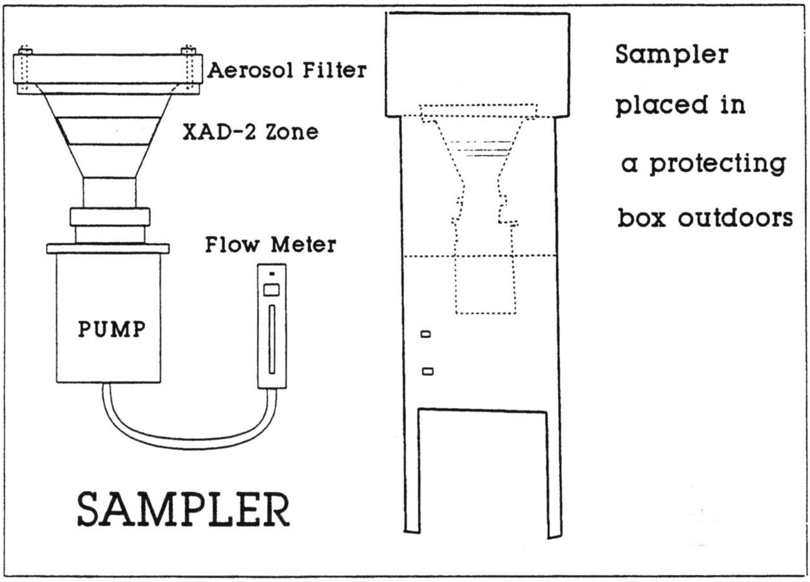


**Figure 5.2.3** Surface high-volume air sampler. An aerosol filter HV3 is placed above a XAD-2 resin bed that collects gaseous organic trace chemicals.[8]

well-protected and trained persons can sample and detect toxins in air by using a bicycle pump with reversed flow and small adsorption tubes with reagents for detection and similar tubes for sampling.

The low-volume sampling methods described above are suitable in studies of short-term problems like odors and accidents. However, in surveys of air pollution and its regular control, high sample volumes are generally taken. Examples of devices used in high-volume air sampling are shown in Figures 5.2.3 and 5.2.4.

## 5.2.6 Fallout Samples

Fallout samples consist of solid particles (dry deposition) and of liquids (wet deposition, snow, organic liquids).

In the case of chemical hazard situations, fallout can be sampled by plastic sheets (for dust and water) and aluminum foil (for organic liquids; e.g., one-square-meter pieces). The sheets are installed in advance on soil and other surfaces at the risk areas. Dry deposition is brushed and liquid pipetted from the surface to a glass container (pipette also to the container),

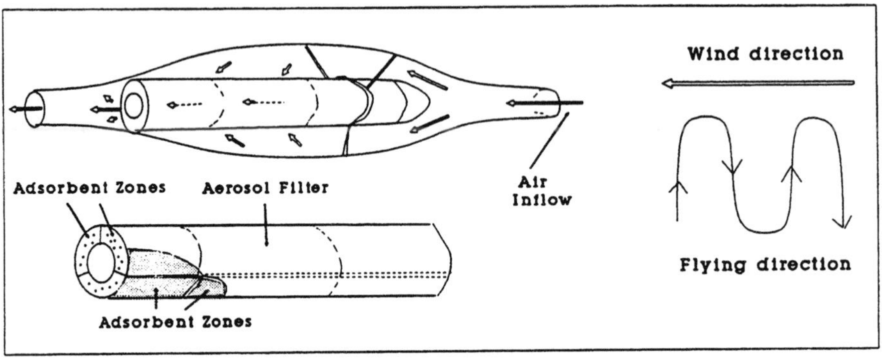


**Figure 5.2.4** High-volume air sampler used in an airplane.[8]

which is closed tightly. From other smooth surfaces, fallout can be collected by wiping with a soft paper, cotton, or glass wool tampon, which can be impregnated with water, ethanol, hexane, etc., depending on the quality of the deposition. After wiping, the tampon is sealed in a glass container or wrapped in aluminum foil.

Long-term fallout sampling is done with standardized devices (Figure 5.2.5). Plastic material is not good for the container device if trace organic compounds are to be studied. In such a case, a glass bottle must be used.

Snow fallout samples are preferably collected from lake or sea ice or other places where snow has been falling on relatively inert surfaces and the time period of fallout is known. A premeasured plastic cylinder (inner diameter known) can be used to collect a snow column. The height of the column should be measured during each sampling. Several columns can be taken from the same sample location and composited in the same clean container to obtain a sufficient sample size. After weighing the sample, one can calculate the fallout as kg of water per square meter during the snowing period sampled. After the snow melts, water can be filtered to form separate solid and liquid (water) samples for analyses.[9]

### 5.2.7 Water Samples

In chemical hazard situations (accidents, chemical warfare), water samples are collected in glass bottles with a sampling scoop by skimming such surface areas where flour, drops, or

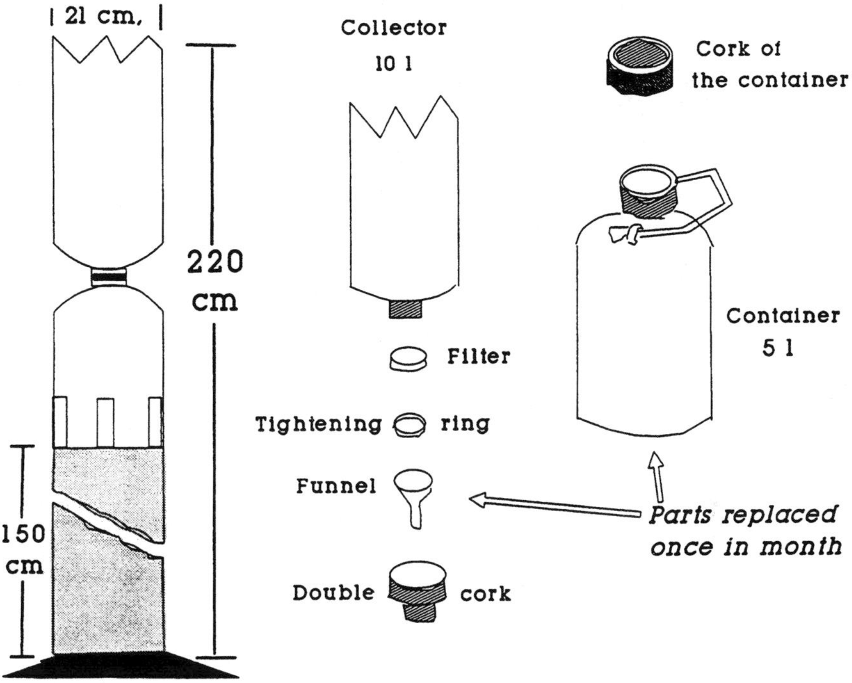


**Figure 5.2.5** Device for rain fallout sampling according to Finnish standard SFS 3865.

oily film are visible. If dissolved toxins are suspected, normal water sampling methods (see below) are used.

For sampling of water from different depths, a simple open bottle or bottle from which the stopper is drawn by a rope at the sampling point can be used, but it is better to use the twin-bottle sampler (Figure 5.2.6).

In professional studies on water pollution, however, more accuracy related to the sampling point and depth is needed. In this case, specially designed devices (Figure 5.2.7) must be used.

The material of the sampling devices and containers has to be selected depending on the requirements of the analyses. For inorganic determinations, the collected water sample can be transferred for storage and transport to a paraffined board box such as a PURE PAK milk case. However, this is not good for cyanides or trace organic compounds; a glass bottle is preferred for them. Also, glass adsorbs hydrophobic compounds (for example, polychlorinated biphenyls [PCBs]) from water. The

**Figure 5.2.6** Twin-bottle water sampler. After the tubing is released from top of the longer glass tube, water flows through the smaller bottle to the larger. Then the water sample in the smaller bottle is very representative.

adsorbed organic compounds can, however, be washed back to analysis with organic solvents. It is recommended to have some milliliters of the extraction solution, e.g., pure hexane, in the sample container before adding the water sample.

A sampling method for surface water that avoids cross-contamination is suction through adsorption tubes similar to those used for air sampling, as described in Figure 5.2.2. The volume of the sample is measured in a suction flask equipped with a measuring scale placed after adsorption tubes but before the pump.

### 5.2.8 Oil Samples

In cases of serious oil accidents, oil samples are collected from surfaces. They are often obtained as mixtures with water which, in storage, causes degradation of oil. Therefore, the water should be removed to 3% or less (w/w) contents.

In monitoring small, invisible oil pollution by water analysis,

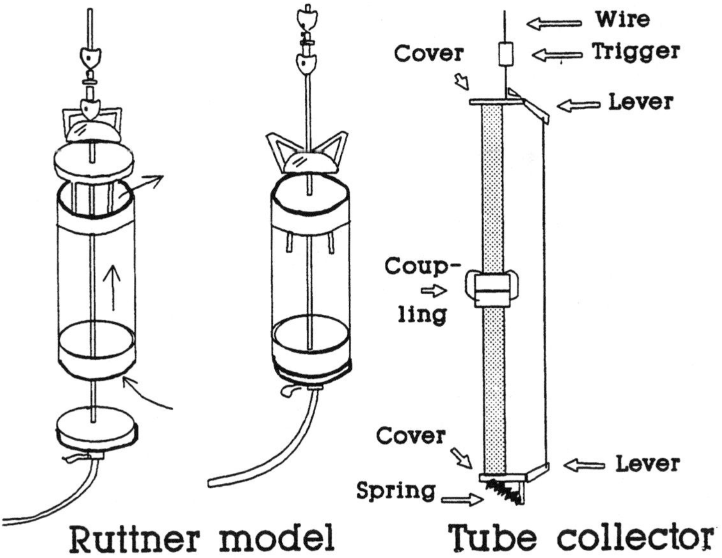


**Figure 5.2.7** Water sampling devices.

the same devices as generally used for water sampling (see above) can readily be used. Immediate addition of a biocide like 0.01% mercury-(II)-chloride is necessary. This prevents biodegradation of the oil traces by bacteria. The water sample must be transferred as fast as possible from the sampler device to a glass bottle containing the extraction solvent (dichloromethane or carbon tetrachloride). An alternative water sampling method for oil trace analyses is suction through an adsorption tube as described in Section 5.2.7.

## 5.2.9 Soil and Sediment Sampling

In cases such as chemical accidents, surface soil can be shoveled into sample containers. Some surface-sediment collecting devices resemble tools for house or road-construction work. Others have traps drawn along the bottom (e.g., grab samplers) or dry-ice cooled plates (Figure 5.2.8).

The sampling location of bottom sediment must be selected from even areas or pits in the bottom where sediment is deposited evenly. On the tops and slopes, the material runs unevenly, and consequently nonrepresentative or no sediment samples are obtained there.

Soil and sediment layers are sampled with corers (Figure

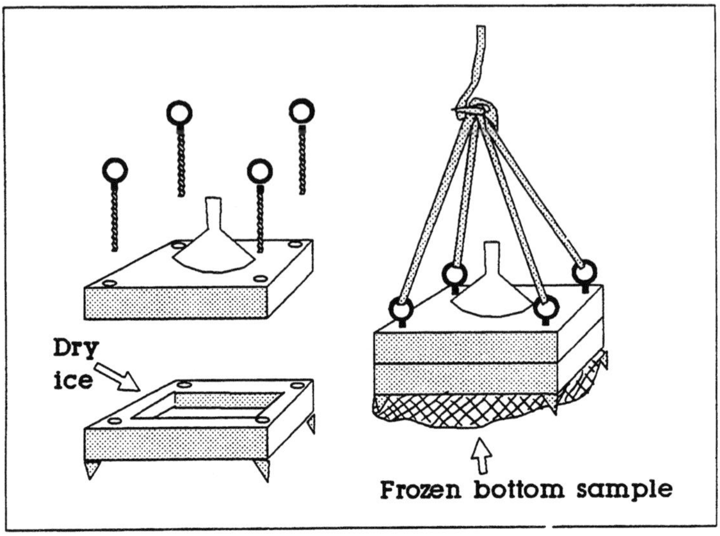


**Figure 5.2.8** Cold-plate sampler for surface sediment developed by J.Särkkä in the University of Jyväskylä, Finland.

5.2.9). The sampled column (core) can be sliced to pieces having a measured height before packing for transport and storage. Soil and peat slices are wrapped in aluminum foil. The wet sediment slices need glass containers. The samples are preferably stored deep-frozen. Soil and sediment layers can be dated with the aid of biostratigraphy[10,11] or by radioisotope measurements.[11,12]

Several tube sampling devices used with standardized procedures have been developed for soil-layer sampling.[1,2] The same methods used for sediments are generally valid for dating of the soil cores.

## 5.2.10 Biological Samples

For monitoring persistent organic compounds in the environment, biota representing different trophic levels are sampled. Widely distributed common species are recommended in order to evaluate useful comparisons between localities, areas, and time periods. During sampling, sex, age, length, and weight of the specimens should be determined and marked on the label or documents which follow the sample. As a rule, the sample is

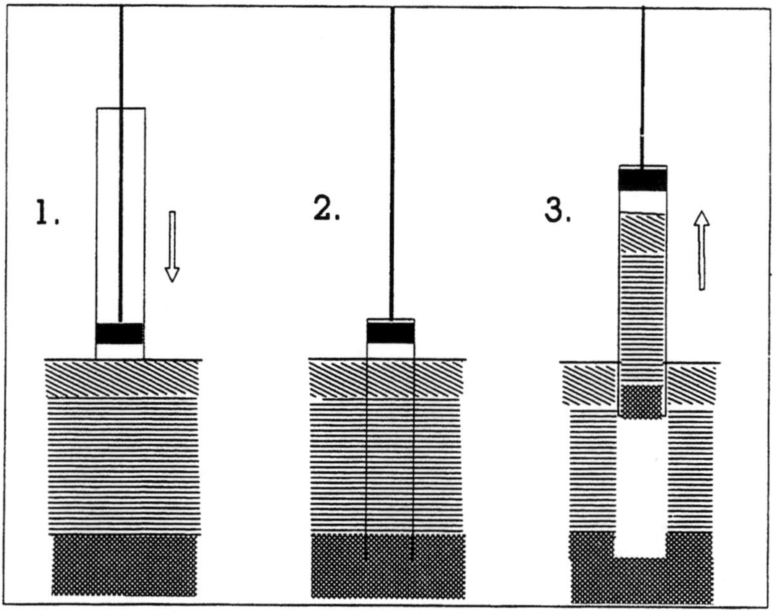


**Figure 5.2.9** Sediment sampling by piston corer.

first wrapped in aluminum foil and then packed in paper, plastic, or any type of box.

In chemical hazard situations, safety is more important than any exact data on species. The sample should be placed in a glass jar with a tight screw cap and packed in a hard box with soft adsorbing filling.

For trace-metal analyses, metal-free sampling tools, e.g., Teflon-coated or clean plastic tools, must be used, and the samples can be sealed in acid-cleaned polyethylene bags.

## 5.3. PRESERVATION AND STORAGE OF THE SAMPLES

### 5.3.1 Principles

Degradation of the chemicals during transportation or storage can lead to errors in monitoring results. To minimize these errors, water samples can be preserved with mercury-(II)-chloride or sulfuric acid, and kept at +5°C or, in case of a long storage period, at –18°C. Solid samples including biological samples are, as a rule, stored frozen, preferably deep-frozen at lower than –40°C temperature.

### 5.3.2 Environmental Specimen Bank

The purpose of an environmental specimen bank is to promote monitoring and research of natural environments by collecting, storing, and disseminating information on biological and other specimens from nature. Together with other monitoring activities, the environmental specimen bank can allow observation on trends in the environment, analyses of the causes of recorded variations, and prognoses of future developments.[13]

Success of a specimen bank depends on the feasibility of preservation methods applied. Preliminary studies indicate that some chlorohydrocarbon traces (DDT, DDE, and DDD) in fish samples decreased significantly during ten years of storage at –18°C, while other toxins (PCB, methyl mercury) persisted. More polar compounds such as chlorophenols degraded nearly completely in three years of similar storage of fish.[14]

It has been suggested, instead of fresh tissue storage, to grind the sample with magnesium sulfate and divide this homogenate into pellets that are stored in a freezer. Some indication of better preservation of chlorohydrocarbons in tissue samples by this method has been obtained.[15]

In several industrialized countries, environmental specimen bank activities have been executed by stepwise study projects in which a large number of parallel samples are taken from each population. One fraction of the samples is analyzed immediately, and the others are stored by several different preservation methods. In certain intervals (years, etc.), repetition of the analyses is done. The remaining samples preserved by the best methods form the permanent specimen bank, which could be very useful in a time scale of hundreds of years.

## 5.4 ANALYTICAL METHODS

### 5.4.1 Working Scheme

A general scheme of the analyses of trace contaminants is shown in Figure 5.4.1.

### 5.4.2 Direct Measurements

Environmental chemicals are often measurable directly in the field, especially from water and air with passive and active

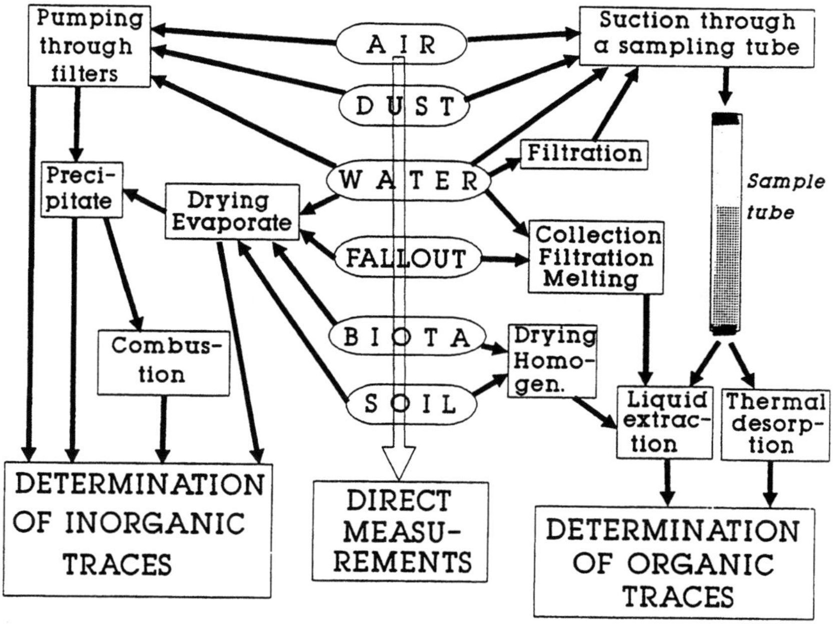


**Figure 5.4.1** General scheme of the trace analyses in environment.

sensors. Examples of directly measuring devices immersed in water are pH-meters, conductometers, and ion-specific electrodes. The gas phase can be sensed by irradiation meters, spectrometers, ionization sensors, electrochemical cells, photoacoustic and chemoacoustic crystals and thermoacoustic devices. Infrared spectrometry is able to detect molecules in the atmosphere (Figure 5.4.2).

Laser devices are useful in remote sensing of pollutants (Figure 5.4.3).[17]

Instead of sampling for laboratory determinations, the common air pollutants can be measured *in situ*. Automatic measuring devices in the field utilize UV-fluorescence for sulfur gases ($SO_2$ and $H_2S$) and chemiluminescence for nitrogen oxides.

## 5.4.3 Sample Cleanup

As a rule, the environmental traces must be isolated from most of the sample matrix before final determination. The first isolation step is most often filtration to separate solid and liquid phases. The soluble traces are isolated from solid or liquid sample or from adsorption tubes by solvent extraction. Volatile

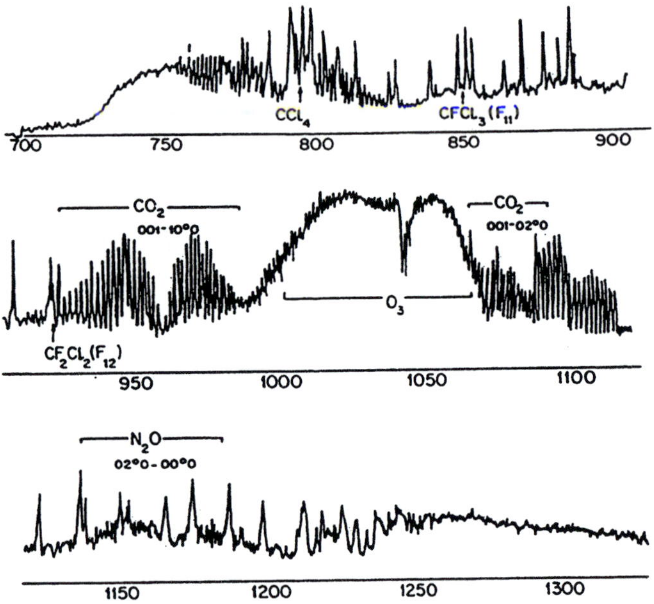


**Figure 5.4.2** Infrared absorption spectrum of the night sky where carbon dioxide, ozone, and nitrogen oxide, as well as the anthropogenic pollutants carbon tetrachloride and freons, were detected.[16]

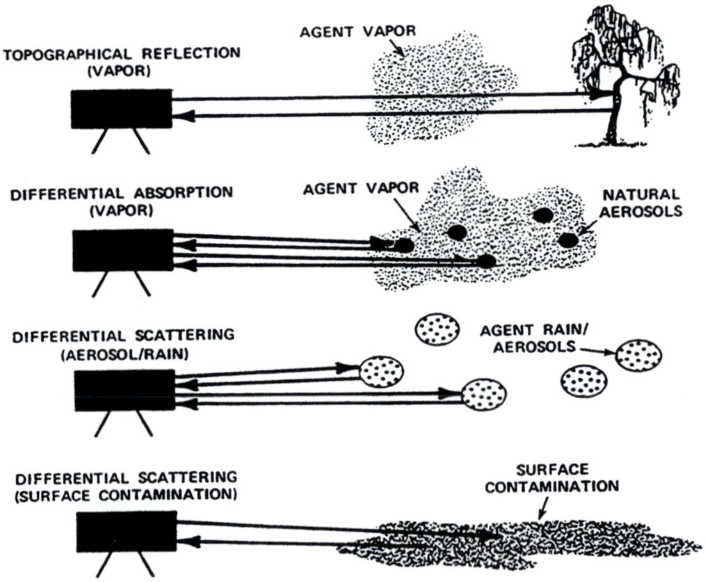


**Figure 5.4.3** Review of laser detection modes.[17]

compounds can be isolated by gas (thermal) desorption from adsorption tube or from solid sample. Also special extractions can be utilized, for example, with pressurized supercritical fluid.

In studies of persistent environmental traces, the biological sample can be degraded, e.g., by digestion with alkali (methanolic KOH) prior to extraction or adsorption.

Often, the extract or desorbed effluent is still too complex for final determination. Further cleanup of the most persistent trace substances (e.g., dioxins or halogenated hydrocarbons) is conveniently done by shaking the extract (hexane solution) with concentrated sulfuric acid that breaks down most of the interfering substances.

Chromatography is the most widely used methodology for further cleanup (group separation) and is also the basis of a large number of final determination methods.[18]

Cleanup (group separation) is very generally done in thin layers (TLC) or columns (column LC) containing adsorbents like alumina or activated coal or chromatographic partition fillings like water on cellulose, polyglycol, or alkylsilanized polyglycol (reversed phase) on silica gel. Organic solvents are most common in elution of the analyzed substances in adsorption or straight-phase partition columns. Reversed-phase columns work best with hydrophilic mixture-eluents such as alcohol-water or acetonitrile-water.

Ion-exchange chromatography is useful in cleanup and final determination of inorganic and organic ions and, in many cases, of amphoteric or polar organic compounds as well.

Ultrafiltration (UF) through membranes, electrophoresis, and size-exclusion chromatography (SEC) separate large molecules. They can be used for isolation of the small-molecular-weight trace substances from persistent macromolecules but also for quantitation and molecular-size distribution determinations of the latter.

An example of a cleanup procedure for persistent organochlorine compounds in biological samples is illustrated in Figure 5.4.4.

### 5.4.4 Final Determinations

Inorganic trace analyses are predominantly elemental determinations that can be done by spectrometry. The most common

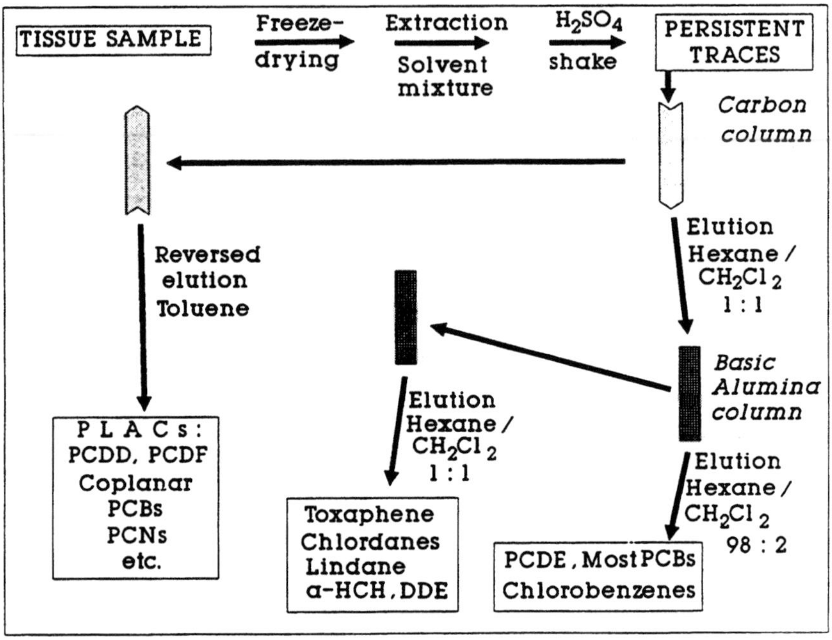


**Figure 5.4.4** Cleanup of persistent organochlorine traces in biological samples. The working procedure was developed in the University of Jyväskylä, Finland. The traces are divided into three fractions for final determinations with gas chromatography coupled to an electron capture detector or a mass spectrometer.

methods used are emission (from flame or plasma) spectrometry, atomic absorption, X-ray fluorescence, mass spectrometry, and activation analysis.

Measurement of organic traces still present after cleanup quite commonly calls for further separation connected to detection devices. The most efficient and convenient method is gas chromatography (GC) with capillary columns, which is to be preferred whenever the trace substances of interest can be eluted by it.

Introduction of the sample to GC after normal cleanup, such as that described in Figure 5.4.4, is done by injecting a microliter amount of a concentrated fraction into the GC injector with split sample flow. There are many types of GC injectors.[18] Some can be used in a "splitless" mode, where the larger part of the injected sample is introduced to the analysis. Dry-needle techniques, where the sample solution is evaporated on the syringe

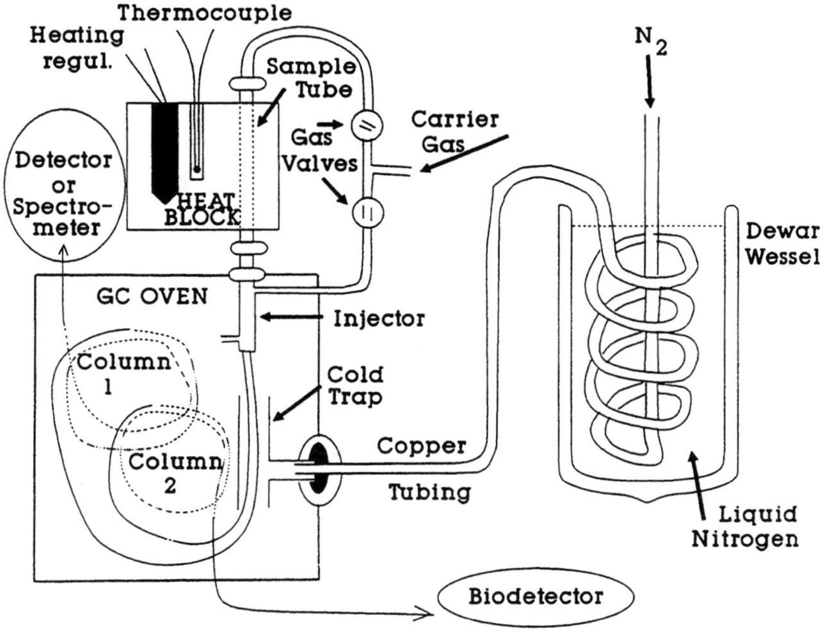


**Figure 5.4.5** Two-column GC equipped with thermal desorption.[19]

needle before splitless injection to improve both sensitivity and resolution of the gas chromatograms are also used.

Some other techniques allow an instant transfer from sampling to GC. Such is the headspace analysis (HSA), where the sample is drawn from the equilibrated gas phase above the matrix straight into the GC. By thermal desorption from an adsorption tube, all volatiles in the sample are transferred to the GC. A cold trap for collecting the sample prior to introduction onto the columns improves the resolution (Figure 5.4.5). After the desorption period, the cooling gas (nitrogen) is switched off. Then the hot oven heats the trapped sample quickly and elution on the GC columns begins.

Twin-column GC using identical columns (Figure 5.4.5) or splitting the effluent from a single column to two detectors allows a simultaneous use of two measurements from the same compound (GC peak). Both detectors can be electronic, but another might also be a biosensor such as a sniff device (Figure 5.4.6). Electronic measurement results (spectrum and/or specific detector signal) together with flavor information from the

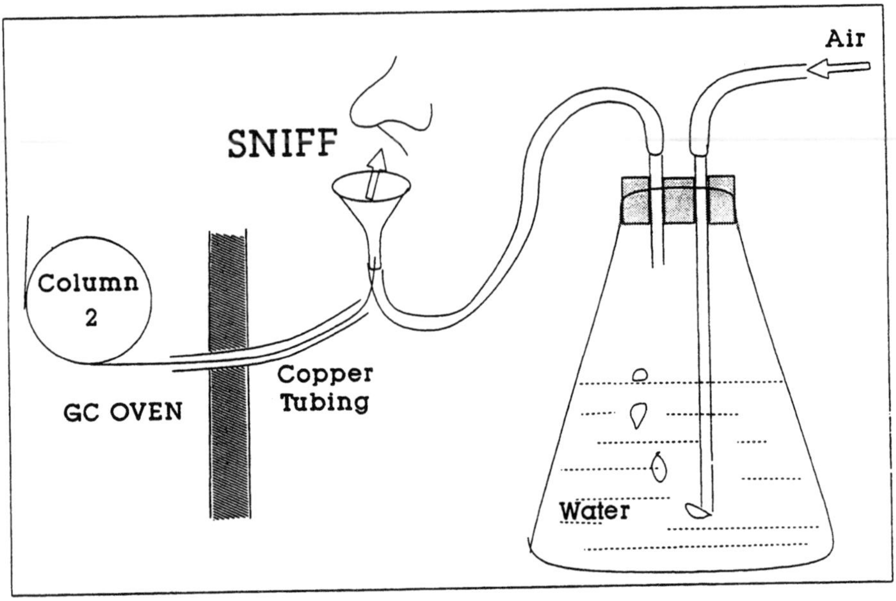


**Figure 5.4.6** Sniff detection coupled to GC.[19]

peak can give the structure of a desired aroma compound or help in identification of a harmful off-flavor substance.[20]

Other biodetections can be achieved by collecting fractions from the GC column outlet, e.g., to a tasteless vegetable oil for taste or toxicity tests. Toxicity or influence upon the behavior of an animal species can be studied on-line by leading (with air blow) the GC effluent to a box where the test individual is kept.

The electronic GC detectors are chosen according to a specific analytical goal. The most general detector is based on thermal conductivity (TCD). The flame ionization detector (FID) is general for carbon compounds. Electron capture detector (ECD) is very sensitive to halogen compounds or compounds containing metals. The alkali thermionic (AT) detector is sensitive to nitrogen and phosphorus, and a flame photometric detector (FPD) is effective with sulphur and phosphorus compounds. The photoionization detector (PID) can be installed to respond selectively to aromatics or certain heteroatoms in organic skeletons. Plasma-generated atomic emission (AE) gives specific signals at different recording channels for large number of elements in the effluent substances.

Some detectors work after catalytic conversion of the eluate.

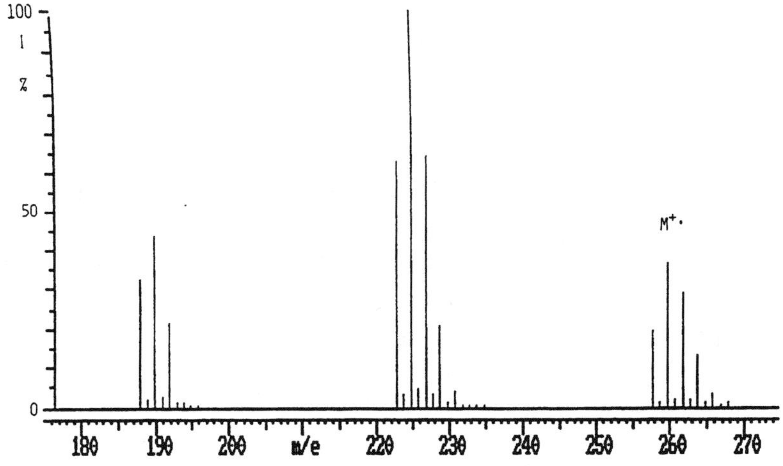


**Figure 5.4.7** Low resolution mass spectrum recorded in a GC/LRMS study of oil residues in Baltic blue mussels (1979). The compound is readily identified with hexachlorobutadiene from characteristic isotope patterns (natural abundances of $^{35}Cl$ and $^{37}Cl$ are 100 and 32.5%, respectively). These results were given as a task to the XX International Chemistry Olympiad in Finland 1988. The number of participants was 104 (4 from each of 26 countries). Fifty-five of them correctly identified the substance.

Of these, coulometric and conductometric (Hall) detectors are linear and highly sensitive to halogen, nitrogen, and sulfur compounds.[18]

The mass spectrometer (MS) is the most sensitive of the multipurpose GC detectors for trace substances. The high-resolution MS (HRMS) makes it possible to obtain molecular formulas of the effluent compounds. The low-resolution MS (LRMS), however, is inexpensive and in most cases sufficient for trace analyses of known compounds. Often, previously unknown traces are also identified with the LRMS (Figure 5.4.7).

More intensity is obtained in the MS by focusing the measurement on the selected ion mode (SIM). SIM of the GC effluent (GC/SIM), also called mass fragmentography, is the most sensitive of the common structure-specific determination methods. In most determinations, low-cost quadrupole or ion trap MSs are sufficient.

This method can be called GC/LRMS/SIM. In pg/g-level analyses, comparison with the authentic model substance by retention time and the intensity ratios of two or three selected ions is the minimum requirement for confident identification. However, errors due to interfering substances may occur. They can be eliminated partly by using HRMS (GC/HRMS/SIM) and, often more reliably, by employing GC/GC techniques.

An effective GC/GC technique is the "heart-cut," where a fraction of the eluate that consists of overlapping peaks is led to another column, e.g., from a nonpolar to a polar column. There the components of the fraction often are readily separated and can be measured by a separate detector.

A typical case of interference that was not solved by GC/HRMS is the separation of all pentachlorobiphenyls (46 possible isomers) from each other. The separation from commercial PCB mixtures is successful by GC/GC/ECD.[21]

Infrared (IR) spectral data are complementary to mass spectral ones. IR can distinguish isomers and ring compounds, which give identical mass spectra. Therefore, a modern Fourier transform infrared (FTIR) spectrometer connected to the GC improves the selectivity in the determination of trace organics. A light pipe (laser) connection in a GC/FTIR/MS instrument can, in structure-specific determinations, reach a limit of determination of 10 ng/g. A matrix isolation technique instead of the light pipe might improve the sensitivity to one tenth.

If GC/FTIR/MS is not sufficient to identify a trace substance, a nuclear magnetic resonance (NMR) spectrum could give an unambiguous solution. However, the sensitivity of NMR is too low for on-line use. The trace peak must be collected from the GC with a cooled trap and dissolved in an NMR solvent. In high field $^1H$ NMR, the limiting factor for sensitivity is not the amount of substance but the solvent impurities. The signal-to-noise ratio improves by time of running, but solvent impurity peaks are increasing also and might overlap the peaks of the substance. In some cases gaseous traces without solvent can be measured. Microgram amounts of low-molecular organic substances are sufficient to obtain reasonable $^1H$ NMR spectra.

The methods described above are useful in screening analyses of the structure of previously unknown and potentially harmful environmental contaminants. Cleanup fractionation can be guided by measurement of the biological activity.

Also, structure-activity relationships (SARs) can be used to guide the fractionation, e.g., in isolation of unknown planar aromatic compounds (PLACs) (Figure 5.4.4) by their strong adsorption to carbon, as first suggested by Jensen and Sundström.[22]

Liquid chromatography (LC) must be used, instead of GC, for determination of compounds that are nonvolatile or degrade in GC and that cannot be transferred to volatile thermostable derivatives. Most common LC detectors utilize UV absorption, fluorescence, and refractory detectors. However, many detectors used in GC are developed for LC, as well. Coupling to MS has been succeeded by several techniques. Nearly as high resolution as in GC can be achieved with the modern high-performance liquid chromatography (HPLC) columns.

An intermediate technique between GC and LC is supercritical fluid chromatography (SFC), where pressurized gas (e.g., $CO_2$) is used as eluent.

## 5.5 EXAMPLES OF CASES

### 5.5.1 Analysis of Air Pollutants

Air pollutants in South Finland were studied by low-volume sampling to Tenax-filled adsorption tubes. Thermal desorption to two identical SE-30 columns leading to ECD and FID produced chromatograms showing aromatic hydrocarbons (FID) and halogenated hydrocarbons (ECD). An example of the results is illustrated in Figure 5.5.1. The sample volume was only 2 liters of air collected in 10 minutes. However, the use of a cold-trap system resulted in high sensitivity. Nanograms of halogenhydrocarbons and micrograms of aromatic hydrocarbons in one cubic meter of air were detectable. Toluene appeared to be the most abundant of the studied pollutants at this semi-urban area near Helsinki.[22]

An example of analysis of organic sulphur gases by GC/FPD and GC/MS is shown in Figure 5.5.2.

### 5.5.2 Chloroanisoles and Veratroles in Fish

Polychlorinated anisoles (PCAs) and veratroles (PCVs) are formed in aquatic environments by biomethylation of chlorinated phenols (PCPs), catechols (PCCs), and guaiacols (PCGs). After discharge of a pulp chlorobleaching mill, they bioaccumulated in the fish to such an extent that the fish became

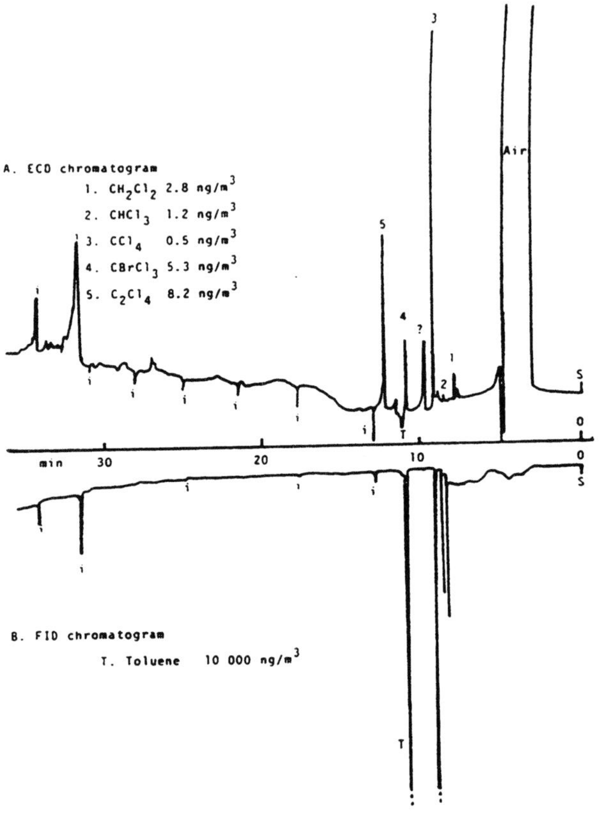


**Figure 5.5.1** Chromatograms obtained by application of the thermal desorption/cold trap system (Figure 5.4.5) to study the ambient air. The columns were two identical quartz capillaries coated with SE-30 phase. The detector on column 1 is ECD and on column 2 FID.[22]

intensively tainted. The bad taste was shown to be caused by PCA and PCV with combined use of instrumental and sensory determinations.[23]

An example of the GC method used for analysis of PCA and PCV in fish is given in Figure 5.5.3. Authentic model substances were used as reference in dual column GC. Measured amounts of 2,3,6-trichloroanisole and 4,5-dibromoveratrole were added to samples prior to the extraction to serve as internal standards for quantitation. No sulfur acid shake could be used in cleanup, because substances to be measured would degrade. The PCA and PCV fraction was isolated from other ECD-active compounds by LC in alumina and silica columns.

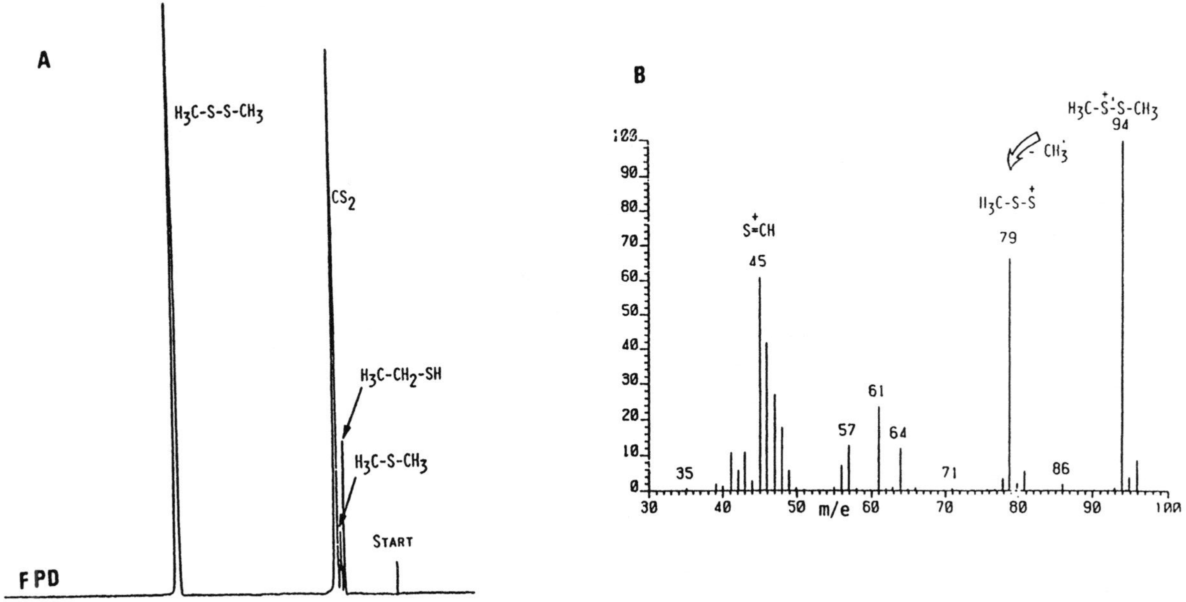


**Figure 5.5.2** Analysis of organic sulfur compounds in working room air. (A) Separation achieved with a glass capillary column measured by a self-built flame photometric detector (FPD). (B) Mass spectrum of the peak of dimethyl disulfide recorded on a Finnigan MAT 212 mass spectrometer (University of Jyväskylä, Finland).

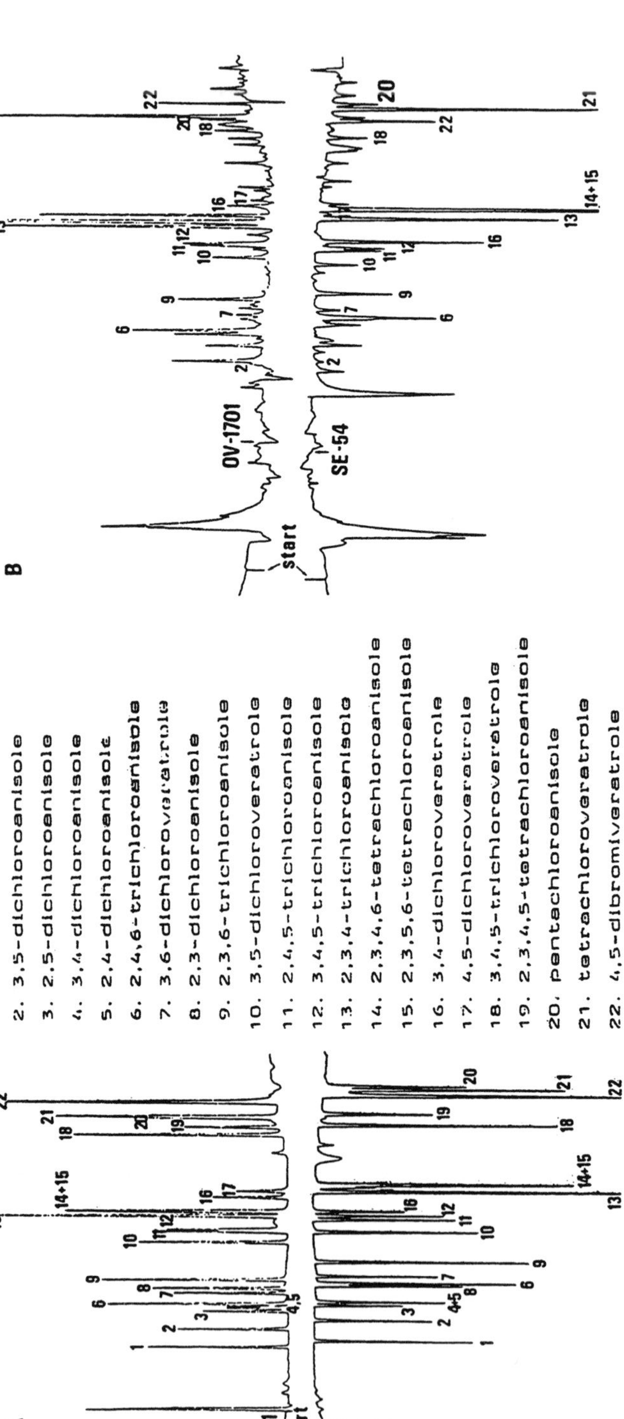


**Figure 5.5.3** Dual-column gas chromatograms. (A) Mixture of model polyhalogenated anisoles and veratroles. (B) Purified extract of lateral muscle of trout exposed to 0.5% kraft pulp mill total effluent in pure lake water. Peaks No. 9 and 22 are from added internal standards.[23]

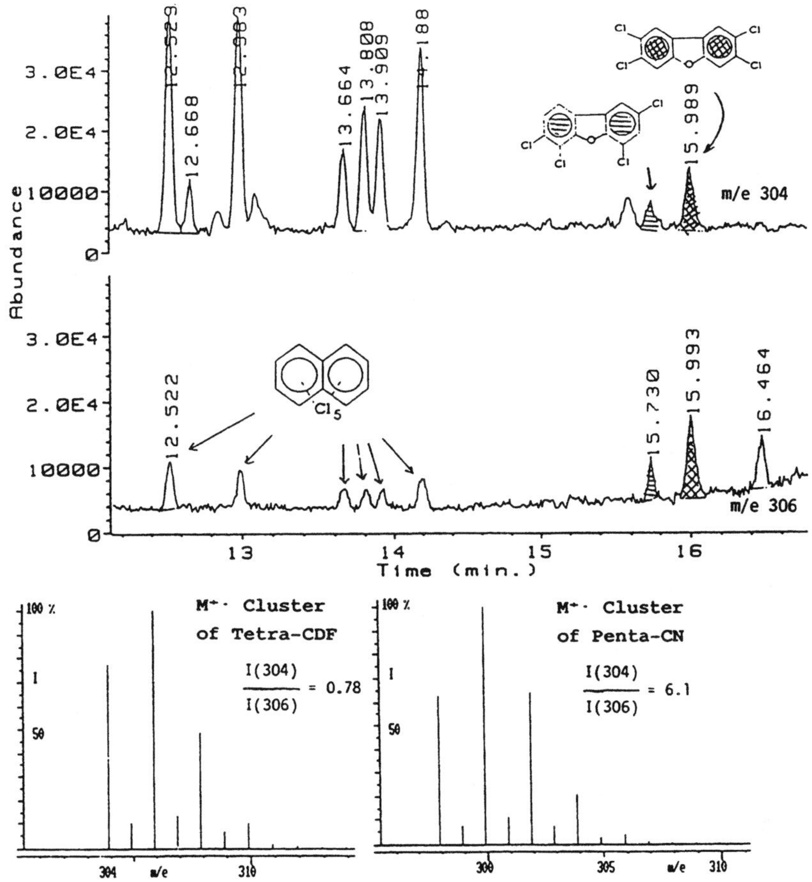


**Figure 5.5.4** Two-ion GC/SIM of the PLAC fraction of a muscle sample of salmon from Baltic proper in 1985. Ion ratio and retention times compared with authentic reference models verified the identity of two TCDF congeners: nontoxic 2,4,6,7-TCDF (10 pg/g) and toxic 2,3,7,8-TCDF (20 pg/g). Six of the other peak pairs can be concluded to be from pentachloronaphthalenes — their ion ratio I(304):I(306) was about 6. Four other interfering compounds appeared at ion channel 304 and one at 306.

## 5.5.3 PLAC Analysis by Mass Fragmentography

GC/LRMS/SIM was used for the determination of dioxin-related toxicants in Baltic salmon.[24,25] The common toxicant 2,3,7,8-TCDF was determined from the PLAC fraction with GC/SIM by recording ions m/e 304 and 306 at the elution time window of tetra-CDF congeners (Figure 5.5.4).

## 5.6 SUMMARY

Sampling and analyses in ecotoxicological studies must be planned together. Direct measurements in environment are possible for some pollutants, but most of them are determined in the laboratory from samples collected from the field. The size and number of the samples from each site are determined by the accuracy needed, the degree of variation in the population studied and the method used, and economical optimization of the study program.

Procedures and instruments for sampling must be prepared to avoid cross-contamination of the samples. Zero samples are taken, handled, and analyzed to monitor the validity of sampling. Special sampling instruments are used in collecting air, precipitation (fallout), water, and sediment samples. Preservation at low temperature is essential for biological samples.

Various spectrometric and electrochemical sensors are used in the direct field measurements of toxicants. The laboratory samples, in most cases, need pretreatments (cleanup) before final determinations. Filtration, solvent extraction, and thermal desorption are typical pretreatments. Chromatographic methods are widely used in both cleanup and final determination. The combination of several detection devices with chromatography and the use of multicolumn techniques improves identification of the pollutants studied.

## 5.7 REFERENCES

1. **Bicking, C. A.** "Sampling," in *Kirk-Othmer. Encyclopedia of Chemical Technology, Vol. 17, 2nd ed.* (New York:Wiley, 1968), pp. 744-762.
2. **Seymour, R. B.** "Sampling," in *Kirk-Othmer. Encyclopedia of Chemical Technology, Vol. 20, 3rd ed.* (New York:Wiley 1968), pp. 525-558.
3. **Stevens, D. J., Jr.** "Field Sampling Design," Chapter 4 in *Ecological Assessment of Hazardous Waste Sites*, W.Warren-Hicks et al., Ed., EPA Document E652/11/21/88-F (1989).

4. **Paasivirta, J., and R.Paukku.** "Use of Composited Samples to Optimize the Monitoring of Environmental Toxins," *Chemosphere* 19:1551-1562, (1989).
5. *Chemical Trends in Wildlife. An International Cooperative Study* (Paris:OECD, 1980).
6. **Paasivirta, J.** "Chlorinated Hydrocarbons in Organisms," *NORDFORSK Miljövårdsserien* 1:239–249 (1983).
7. **Mack, G. A., and P. E. Robinson.** "Use of Composited Samples to Increase the Precision and Probability of Detection of Toxic Chemicals," in *Environmental Applications of Chemometrics*, J. J. Breen and P. E. Robinson, Ed., *ACS Symposium Series* 292:174-183 (1985).
8. **Laitinen, R.** "Sampling and Analysis of Organic Pollutants in Ambient Air," Lic. Phil. Thesis. University of Jyväskylä, Finland (in Finnish). Jyväskylä (1987).
9. **Paasivirta, J., M. Knuutila, and R. Paukku.** "Study of Organochlorine Pollutants in Snow at North Pole and Comparison to the Snow at North, Central and South Finland," *Chemosphere* 14:1741–1748 (1985).
10. **Granberg, K.** "The Diatom Successions in the Recent Sediments and the Euthrophication of Ristiselkä, Lake Päijänne," *Aqua Fennica* 2:20-27 (1972).
11. **Kansanen, P.H., and T. Jaakkola.** "Assessment of Pollution History from Sediment in Lake Vanajavesi, Southern Finland. I. Selection of Representative Profiles, Their Dating and Chemostratigraphy," *Ann.Zool.Fennici* 22:13-55 (1985).
12. **Häsänen, E.** "Dating of Sediments, Based on $^{210}$Po Measurements," *Radiochem. Radioanal. Lett.* 31:207-214 (1977).
13. Nordic Minister Council. Nordic Environmental Specimen Bank. "An Aid to Environmental Monitoring and Research." *NORD B* 1977:6, Oslo (1977).
14. **Paasivirta, J., K. Mäntykoski, M. Knuutila, and J. Koistinen.** "Säily. Studies reported to the Ministry of Environment, Finland" (otherwise nonpublished) I (1985), II (1987), and III (1988).
15. **Mowrer, J., and S. Jensen.** "Evaluation of Current and Proposed Methods for the Long-Term Preservation of Biological Samples for Pollutant Analyses," SNV Rapport 1594, Stockholm (1982).

16. **Hirschfield, T.** "FTIR. Now and in the Future, Part 2," *Eur. Spectrosc. News* 55 (August/September):15-21 (1984).
17. **Carrico, J. P., K. R. Phelps, E. N. Webb, R. A. Mackay, and E. R. Murray.** "Chemical Agent Remote Sensing," Proceedings of the 2nd International Symposium on Protection Against Chemical Warfare Agents, Stockholm, Sweden, 15-19 June (1986), 193-200.
18. **Grob, R. L., Ed.** *Chromatographic Analysis of the Environment*, 2nd ed. (New York:Marcel Dekker, 1983).
19. **Veijanen, A., M. Lahtiperä, R. Paukku, H. Kääriäinen, and J. Paasivirta.** "Recent Development in Analytical Methods for Identification of Off-Flavour Substances in Finnish Freshwater Environments," *Water Sci. Technol.* 15:161-168 (1983).
20. **Veijanen, A., J. Paasivirta, and M. Lahtiperä.** "Structure and Sensory Analyses of Tainting Substances in Finnish Freshwater Environments," *Water Sci. Technol.* 20:43-48 (1988), and Veijanen, A., Dr. Thesis, University of Jyväskylä, Finland (1990).
21. **Schulz, D. E., G. Petrick, and J. C. Duinker.** "Complete Characterization of Polychlorinated Biphenyl Congeners in Commercial Aroclor and Clophen Mixtures by Multidimensional Gas Chromatography-Electron Capture Detection," *Environ. Sci. Technol.* 23:852-859 (1989).
22. **Paasivirta, J., H. Vihonen, J. Salovaara, J. Tarhanen, A. Veijanen, M. Lahtiperä, R. Paukku, E. Kantolahti, and R. Laitinen.** "Application of New Environmental Sampling and Analysis Techniques to Monitoring Chemical Hazards," Proceedings of the International Symposium of Protection Against Chemical Warfare Agents, Stockholm, Sweden, June 1983, FOA Report C 401717-C2,C3, ISSN nr 0347-2124 (1983), 37-44.
23. **Paasivirta, J., P. Klein, M. Knuutila, J. Knuutinen, M. Lahtiperä, R. Paukku, A. Veijanen, L. Welling, M. Vuorinen, and P. J. Vuorinen.** "Chlorinated Anisoles and Veratroles in Fish. Model Compounds. Instrumental and Sensory Determinations," *Chemosphere* 16:1231-1241 (1987).
24. **Paasivirta, J., K. Mäntykoski, J. Koistinen, T. Kuokkanen, E. Mannila, and K. Rissanen.** "Structure Analyses of Planar Polychloroaromatic Compounds in Environment," *Chemosphere* 19:149-154 (1989).

25. **Koistinen, J., J. Paasivirta, and P. J. Vuorinen.** "Dioxins and Other Planar Polychloroaromatic Compounds in Baltic, Finnish and Arctic Fish Samples," *Chemosphere* 19:527-530 (1989).

# 6

# Interpretation of the Environmental Analysis Results

## 6.1 INTRODUCTION

To be utilized, analytical results must be handled variously to reach interpretations and conclusions, depending on the purpose of the study at hand. The requirements of handling of results must be considered when the study is being planned (Figure 6.1.1).

## 6.2 USE OF STATISTICAL METHODS

The results of the environmental analyses produce standard variances, e.g., the population variance estimate:

$$V = \sum_{i=1}^{N} (x_i - \mu)^2 / (N-1) = s^2$$

where $N$ is the number of sample analyses within the population studied, $x_i$ is the concentration of compound analyzed in an individual sample, $\mu$ is the average content, and $s$ is the standard deviation. Values below the limit of determination (DL) are suggested to be taken, instead of zeros, as $x_i = DL/2$ or $DL/3$.

The resulting mean $\mu$ and standard deviation $s$ together with $N$ form a basis for comparison between populations.[1] In the simplest statistical comparisons of means such as $t$-tests, a normal distribution of $x_i$:$s$ is assumed. However, the concentrations of persistent organic trace compounds in human and wildlife populations are frequently not normally distributed; rather, they are lognormally distributed.[2-4] This property is illustrated in Figure 6.2.1.

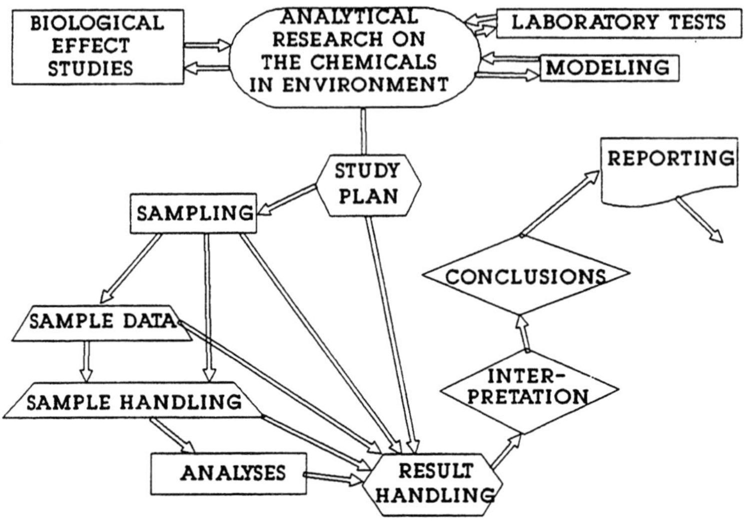


**Figure 6.1.1** Role of result handling and interpretation in a research program on environmental trace chemicals.

Deviation from the normal distribution of concentrations or their logarithms can be expressed as skewness (SK), which is related to the third moment about the mean μ and calculated:

$$SK = \left(\sum (x_i - \mu)^3\right) / (N \times s)^3.$$

If SK is closer to zero for the values of log(concentration) than for concentration, then the distribution of the concentrations is more lognormal than normal. In those cases, standard tests (*t*-test, etc.) for the significance of differences between populations are more reliably done from μ, *s*, and *N* of the log(*c*) values.[2-4]

Earlier, observations with high concentrations of environmental toxicants in biological specimens were deleted from the statistics as "outliers."[5] This procedure was not correct, because these high values actually belong to the underlying distribution and therefore should not be excluded from data handling.

A number of standard statistical calculations are available for handling the analysis results in the environment. At present, these calculations are conveniently done with microcompu-

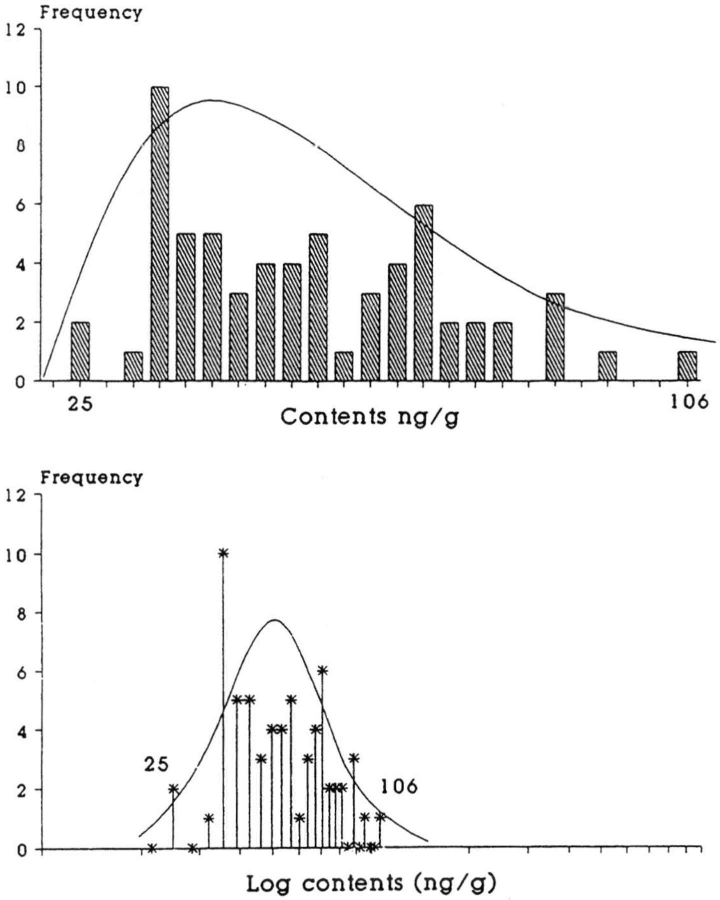


**Figure 6.2.1** Histograms of the concentrations of polychlorinated biphenyls (PCBs) in herring from Gulf of Finland, 1987.[4] Results of 65 determinations from the same population (single catch) clearly indicate that the underlying distribution of the concentrations is lognormal.

ters.[6] Linear correlations suggest which variables depend on each other. For example, the strong positive, significant correlation between concentrations of mercury and weight of individual fish results from the high food-chain enrichment and slow excretion rates of methyl mercury. The significant positive correlation between the concentrations of two trace chemicals points out that both might have a common source.

Analysis of variance (ANOVA) with covariates eliminated (Covariance-ANOVA) together with the Q-test is a powerful treatment to detect significant differences between populations.[1] Regression analysis, also preferably with covariates eliminated, is a good tool in monitoring time trends. A procedure of stepwise multiple linear regression allows determination of the significance of the time period (year) as an explanatory variable. This gives the significance of the time trend, often an important task in environmental monitoring programs.[6]

Other multivariate analyses that belong to the discipline of chemometrics can be applied to the interpretation of complex result data in screening, surveying, and monitoring studies of environmental pollution.[2]

## 6.3 EXAMPLE: MUSSEL INCUBATIONS

Four-week incubation of mussels followed by analyses was used in Finland as a sensitive method for monitoring chlorohydrocarbons, chlorophenols, and aromatic chloroethers in watercourses.[7] Use of composited samples (see Section 5.2.4) was applied in these studies. It proved to give a usefully low variance with five analyses of composites of three individuals (15 mussels) from each sampling station.[7]

In 1984 to 1987, the method was used at some locations in the Lake Päijänne watercourse, but in 1988 the method was applied at 40 sites in Finland to study the influences of the pulp mills and to detect possible chlorophenol and PCB leakages. This time only three composites of five individuals were analyzed from each station.[8]

Eight chlorophenol compounds were found to bioaccumulate quite strongly in mussels. Their formulas and name abbreviations are given in Figure 6.3.1.

Correlations of the concentrations measured from total material in 1988 (40 incubation stations, 119 determinations) are shown in Table 6.3.1.

TeCP and PeCP, which correlate very significantly with each other, are wood preservatives and combustion products. They obviously come from the same sources (leakages and stacks) all around the country. The other group of strongly correlating residues, polychloroguaiacols 345TCG, 456TCG, TeCG, and DMP (trichlorosyringol), are main chlorophenolics formed in chlorobleaching of pulp. Obviously, pulp mills are their main source at the studied water areas.

To study whether process changes of one pulp mill in Central Finland had any influence on chloroguaiacol levels in the

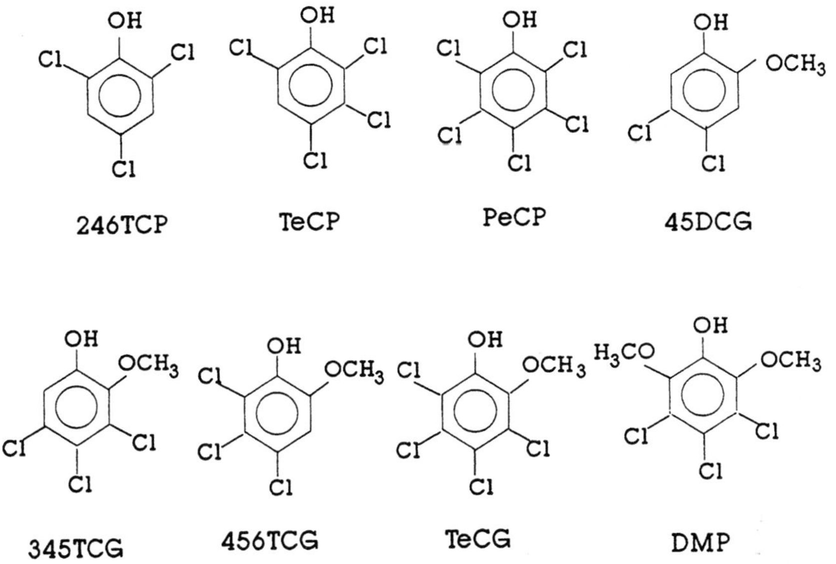


**Figure 6.3.1** Eight chlorophenol compounds that bioaccumulate to incubated mussels.[7,8]

**Table 6.3.1 Correlation coefficients in case mussels, summer, 1988 ($N = 119$)**

| | TeCP | PeCP | 24DCP | 45DCG | 345TCG | 456TCG | TeCG | DMP |
|---|---|---|---|---|---|---|---|---|
| 246TCP | 0.106 — | −0.045 — | 0.482 *** | −0.045 — | 0.447 *** | 0.377 *** | 0.337 *** | 0.083 — |
| TeCP | | 0.497 *** | −0.038 — | −0.090 — | −0.111 — | −0.099 — | −0.106 — | −0.070 — |
| PeCP | | | −0.072 — | −0.095 — | −0.168 * | −0.158 O | −0.169 * | −0.106 — |
| 24DCP | | | | −0.055 — | 0.078 — | 0.143 O | 0.051 — | −0.096 — |
| 45DCG | | | | | 0.159 O | 0.113 — | 0.093 — | 0.071 — |
| 345TCG | | | | | | 0.622 *** | 0.951 *** | 0.371 *** |
| 456TCG | | | | | | | 0.561 *** | 0.614 *** |
| TeCG | | | | | | | | 0.376 *** |

Note: Significances are shown as symbols associated with the probability ($p$) of rejection of the null hypothesis which corresponds the exact linear relationship between two variables. ***: $p =< 0.005$; **: $p =< 0.01$; *: $p =< 0.05$; O: $p =< 0.1$; –: $p > 0.1$

recipient watercourse, mussel incubations were done every summer during 1984 to 1988 in the same sample station (Kuusaankoski).[7,8] For trend detection, multiple linear regression analyses were run with the sum of the concentrations of three compounds:

$$\text{STTeCG} = 345\text{TCG} + 456\text{TCG} + \text{TeCG}.$$

The output of the run with six explanatory variables was

**Mussels Incubated in Kuusaankoski, 1984–1988 (*N* = 29)**

| | | |
|---|---|---|
| Dependent variable | *Y* = STTeCGng/g | Mean = 215.7241 |
| Independent variable | X(1) = Y E A R a | Mean = 85.5517 |
| Independent variable | X(2) = Length cm | Mean = 7.6083 |
| Independent variable | X(3) = A G E yrs | Mean = 7.6272 |
| Independent variable | X(4) = Weight g | Mean = 20.6634 |
| Independent variable | X(5) = Dry tissue g | Mean = 1.0937 |
| Independent variable | X(6) = Fat percent | Mean = 6.4393 |

**Results of multiple linear regression *Y* = Aa + *B*(*i*) * *X* (*i*)**

| Coefficient | Estimate | St. Error | F | P | Signif. |
|---|---|---|---|---|---|
| Aa | 4891.9980 | | | | |
| B(1) | –61.4441 | 29.0172 | 4.4838 | < 0.05 | * |
| B(2) | 131.7315 | 41.4228 | 10.1135 | < 0.01 | ** |
| B(3) | 4.1272 | 136.2008 | 0.0009 | | — |
| B(4) | –18.9893 | 1.3041 | 212.0420 | < 0.001 | *** |
| B(5) | –70.7400 | 19.1342 | 13.6681 | < 0.01 | ** |
| B(6) | 2.5466 | 16.0185 | 0.0253 | | — |

| Source | SS | DF | MS |
|---|---|---|---|
| Regression | 332748.0 | 6 | 55458.0 |
| Residual | 692000.0 | 22 | 31454.6 |
| Total | 1024748.0 | 28 | |

Significance of the time trend could be read from the *F* value of YEAR [row B(1)] = 4.48. Values chosen for reading the *F*-table were V1 = 1 and V2 = 22 (residual degrees of freedom). Consequently, the detected significance for the decreasing trend [$B(1) = -61.4$] was $p < 0.05$ (*; only slight).

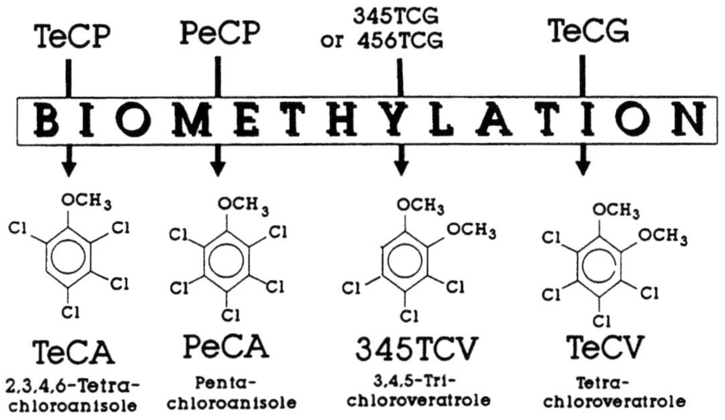


**Figure 6.3.2** Biotransfer of chlorophenolics (formula in Figure 6.3.1) to chloroanisoles and chloroveratroles.

In 1987, mussels were incubated at ten sampling stations of Central Finland for the purpose of surveying a PCB leakage and the influence of a pulp mill on chlorophenol compound contamination.[7]

Chlorophenols are biomethylated in the environment to more lipophilic and persistent chloroanisoles and veratroles (Figure 6.3.2). The latter were also studied in mussels utilizing the analysis method developed for them (see Section 5.5.2).

Sampling stations of the summer 1987 study are shown in Figure 6.3.3. To compare chlorophenol and chloroanisole/veratrole contamination upstream and downstream from a new large pulp mill, results from six stations were selected to ANOVA-Q-test. Station 1 (Matilanvirta) was used as the reference. The station 6 area was polluted by a pulp mill that had been closed in 1985. Station 7 was 0.3 km and station 8 was 2 km downstream from the discharge point of the activated sludge treatment plant of the new pulp mill (started in 1985). Station 9 (Kuusaankoski) was 20 km and station 10 (Torronselkä) was 37 km downstream from the discharge point.[7]

Five composites of three individuals were analyzed from each station. Average concentrations of six chlorophenols (PCPs) and four chloroanisoles/-veratroles (PCAs/PCVs) are shown in Figures 6.3.4 and 6.3.5, respectively.

Contents of ten different chlorophenols, -anisoles, and -veratroles were tested with a combination of ANOVA and the Q-test technique of the Newman-Keuls procedure.[1] If the

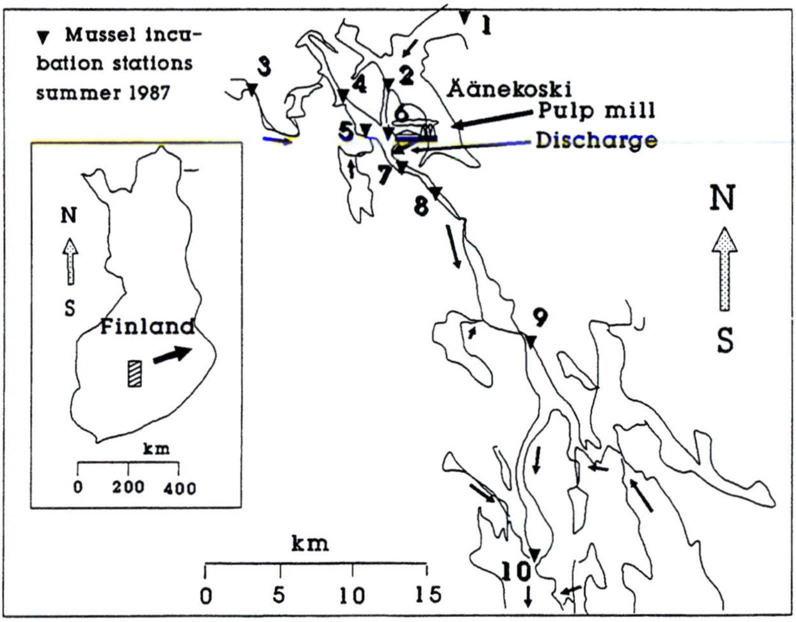


**Figure 6.3.3** Sampling stations for mussel incubation in summer 1987.

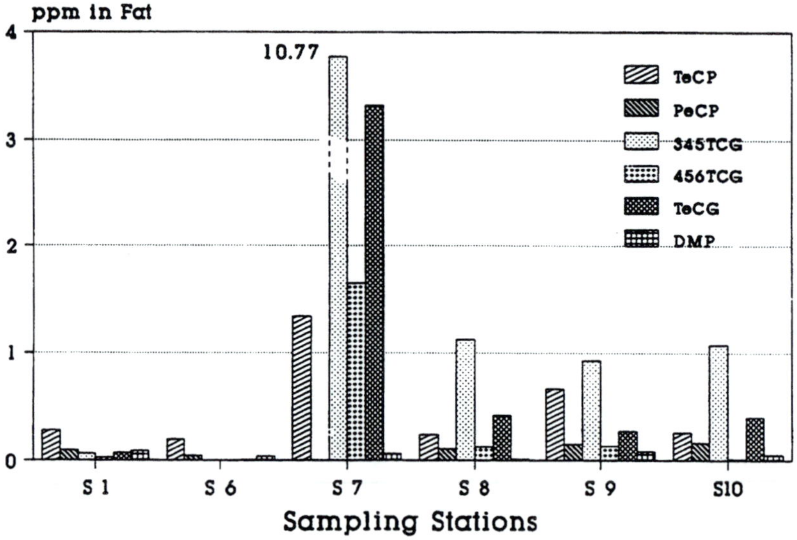


**Figure 6.3.4** Average concentrations in μg/g (ppm) in fat of chlorophenols in mussels incubated in summer 1987.[7]

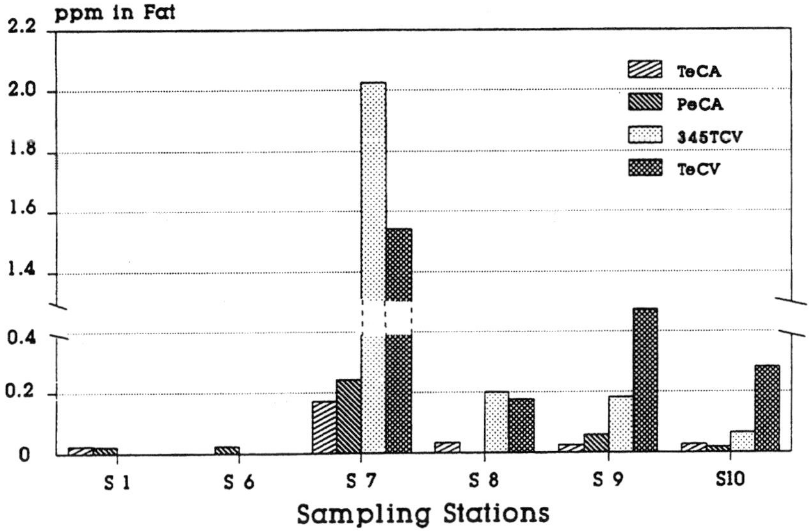


**Figure 6.3.5** Average concentrations in µg/g (ppm) in fat of chloroanisoles and -veratroles in mussels incubated in summer 1987.[7]

ANOVA test results in a significant variance ratio ($F$), the Q-test estimates the significance between populations. An example of the output of microcomputerized ANOVA-Q-test is given below (lipid covariate is eliminated by using concentrations at lipid basis):

**Mussels incubated in 1987** for Q test nr of treatments a is 6 and degrees of freedom v are 29 and standard deviation (SD) is 0.112

ANOVA for variable T e C V

| SOURCE | SS | DF | MS |
|---|---|---|---|
| Populations | 8.462 | 5 | 1.692 |
| Residual | 3.604 | 24 | 0.150 |
| Total | 12.066 | 29 | |

F(5,24) = 11.271 ***

| Population | Mean | SD |
|---|---|---|
| STAT1 | 0.00000 | 0.46037 |
| STAT6 | 0.00000 | 0.46037 |
| STAT7 | 1.54140 | 1.53600 |
| STAT8 | 0.17420 | 0.31081 |
| STAT9 | 0.47300 | 0.29359 |
| STA10 | 0.28100 | 0.22491 |

Overall mean = 0.41160
Overall st. error = 0.17330

The extreme difference of the means is 1.541

Q(a,v) from table = 4.32

Least Significant Difference is LSD = Q * SD = 0.484

LIST OF THE DIFFERENCES:

| | | |
|---|---|---|
| STAT1–STAT6 = | 0.000 | – |
| STAT1–STAT7 = | –1.541 | * |
| STAT1–STAT8 = | –0.174 | – |
| STAT1–STAT9 = | –0.473 | – |
| STAT1–STA10 = | –0.281 | – |
| STAT6–STAT7 = | –1.541 | * |
| STAT6–STAT8 = | –0.174 | – |
| STAT6–STAT9 = | –0.473 | – |
| STAT6–STA10 = | –0.281 | – |
| STAT7–STAT8 = | 1.367 | * |
| STAT7–STAT9 = | 1.068 | * |
| STAT7–STA10 = | 1.260 | * |
| STAT8–STAT9 = | –0.299 | – |
| STAT8–STA10 = | –0.107 | – |

Results of ANOVAQs for mussels are summarized below.

MATRIX of significant (*) and nonsignificant (—) differences

| Station | 6 | 7 | 8 | 9 | 10 |
|---|---|---|---|---|---|
| 1 | — | * | — | — | — |
| 6 | | * | — | — | — |
| 7 | | | * | * | * |
| 8 | | | | — | — |
| 9 | | | | | — |

For TeCP, PeCP, and DMP, *F* was nonsignificant (—); for 345TCG, 456 TCG, TeCG, TeCA, PeCA, 345 TCV, and TeCV, *F* was very significant (***). The Q-test gave the same significance matrix shown here for all seven compounds with very significant *F* (***).

The ANOVAQ procedure is very conservative. In this case, it gave significant differences only between station 7 and others for seven compounds. TeCP and PeCP, which are the main chlorophenol pollutants from wood preservation and combustion, indicated no significant differences. To the contrary, their biomethylation products TeCA and PeCA appeared at significantly higher levels in mussels from station 7. One explanation is that TeCP and PeCP in the discharge from the mill bioaccumulated much less in mussels than their metabolites TeCA and PeCA and did not reach significant concentrations compared to other areas.

The bioconcentration factor (Bcf) of chlorophenols in natural conditions was measured by fish studies between stations 8 and 9:[9]

$$\mathrm{Bcf} = \frac{\text{Contents in fish muscle}}{\text{Contents in the lake water}}.$$

The observed Bcf values in the roach were Bcf = 60 for TeCP, 200 for TeCP, and 1600 for 456TCG. Consequently, the bioaccumulation of TeCP and PeCP is weak, but that of chloroguaiacols stronger. The content of chloroanisoles and veratroles in lake water was below the limit of determination (1 ng/l). The content in fish was 10,000 to 100,000 ng/kg;[10] therefore, bioaccumulation (Bcf values) of polychlorinated anisoles (PCAs) and polychlorinated veratroles (PCVs) could be larger than 100,000.

## 6.4 BIOCIDES IN AQUATIC BIRDS

In the previous section, a method to estimate bioaccumulation by measuring Bcf in natural conditions was described. A very important task is to measure biomagnification. In the case of aquatic (fish-eating) birds, enrichment was studied by comparison of biocides in eggs, juveniles, and adults (Figure 6.4.1).[11]

Content comparison gave no direct measure for enrichment in the local food chain where the juveniles were eating fish (vendace) from the lake. Eggs laid by the mothers contained biocides, and juveniles had an even lower content than eggs. However, while the total biomass of the individuals was known, one could calculate the average total amount of each biocide in individual specimens. Then, in most cases, the amount in juveniles [$A$(juv)] was higher than amount in egg [$A$(egg)]. This was firm proof that the pollutant did bioaccumulate in the juvenile from its food. Comparison of the enrichment power between DDE, PCB and MetHG cannot be done correctly by calculating $A(\text{juv})/A(\text{egg})$, because their content in vendace was very different. In contrast, the relative enrichment factors (Erf) based on increase of the amount in juvenile compared to the concentration in food could be used:

$$\text{Erf} = \frac{A(\text{juv}) - A(\text{egg})}{C(\text{ven})}.$$

The results of Erf calculation are shown in Table 6.4.1. The highest enrichment factors were measured in black-headed gull samples (4.55 for DDE, 6.41 for PCB, and 1.06 for methyl mercury). This indicates that the persistent chlorohydrocarbons are enriched to higher levels than methyl mercury.

This result is in agreement with results from Baltic white-tailed eagles, which represent the highest trophic level of the

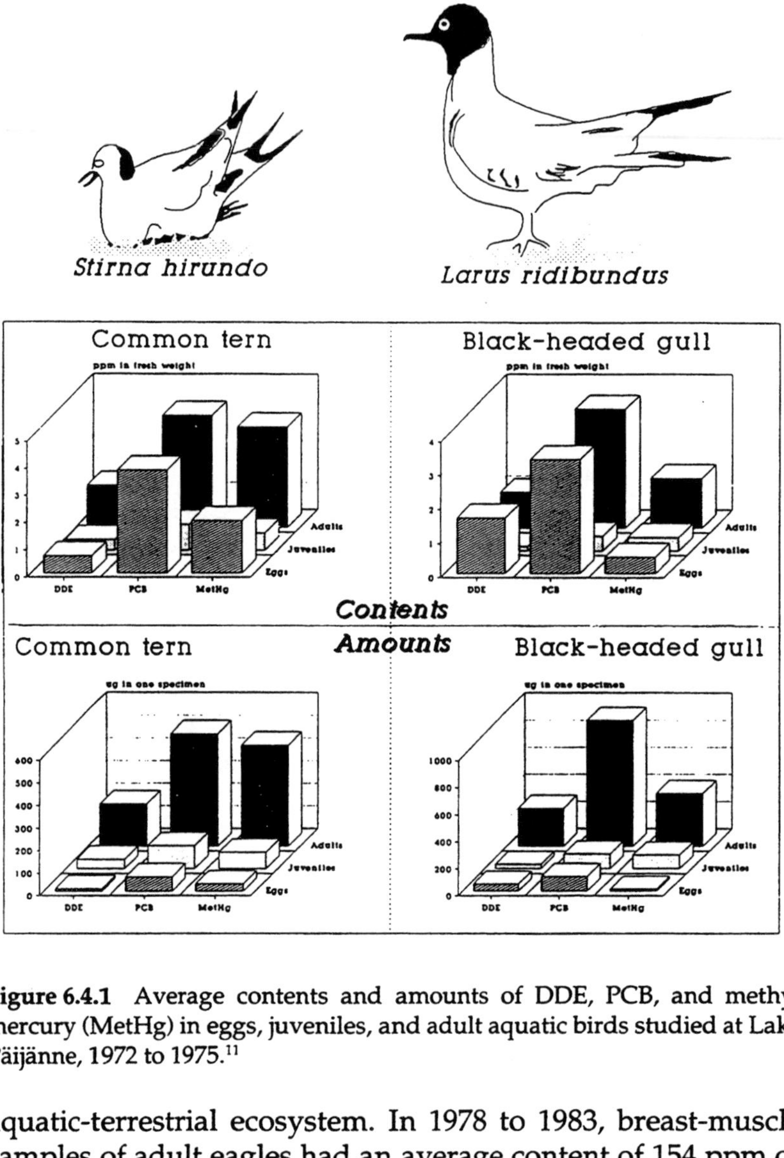


**Figure 6.4.1** Average contents and amounts of DDE, PCB, and methyl mercury (MetHg) in eggs, juveniles, and adult aquatic birds studied at Lake Päijänne, 1972 to 1975.[11]

aquatic-terrestrial ecosystem. In 1978 to 1983, breast-muscle samples of adult eagles had an average content of 154 ppm of DDE, 310 ppm of PCB, and 7.95 ppm of methyl mercury. If these averages are divided by the average contents in juvenile starlings (which have a low position in the food chain) from South Finland measured in the same time period, the following enrichment ratios (ERR) are obtained:[12]

$$ERR = \frac{C(\text{eagle})}{C(\text{starling})}.$$

**Table 6.4.1 Amounts μg/specimen, concentration in fresh fish (vendace) μg/kg, and relative enrichment factors of three biocides in Lake Päijänne birds 1972 to 1975[11]**

| Species | Biocide | *A*(juv) | *A*(egg) | *C*(ven) | Erf |
|---|---|---|---|---|---|
| Great crested grebe | DDE | 138 | 234 | 42 | [a] |
| *(Podiceps cristatus L.)* | PCB | 466 | 204 | 117 | 2.24 |
| | MetHg | 266 | 30 | 415 | 0.57 |
| Common tern | DDE | 42 | 11 | 42 | 0.74 |
| *(Sterna hirundo L.)* | PCB | 103 | 64 | 117 | 0.33 |
| | MetHg | 73 | 33 | 415 | 0.10 |
| Black-headed gull | DDE | 33 | 53 | 42 | [a] |
| *(Larus ridibundus L.)* | PCB | 110 | 109 | 117 | 0.01 |
| | MetHg | 103 | 16 | 415 | 0.21 |
| Common gull | DDE | 85 | 176 | 42 | [a] |
| *(Larus canus L.)* | PCB | 144 | 218 | 117 | [a] |
| | MetHg | 350 | 20 | 415 | 0.80 |
| Herring gull | DDE | 712 | 668 | 42 | 1.05 |
| *(Larus argentatus L.)* | PCB | 1525 | 930 | 117 | 5.09 |
| | MetHg | 523 | 51 | 415 | 1.14 |
| Black-backed gull | DDE | 863 | 672 | 42 | 4.55 |
| *(Larus fuscus L.)* | PCB | 1203 | 453 | 117 | 641 |
| | MetHg | 496 | 56 | 415 | 1.06 |

[a] No enrichment: juveniles excrete more than they collect.

For DDE, ERR = 4700; for PCB, ERR = 3600; and for MetHg, ERR = 88.

For other biocides, the following ratios were obtained:[12] for lindane, ERR = 46.3; for hexachlorobenzene (HCBz), ERR = 295; and for the sum of chlordanes (SCHL), ERR = 5000.

Accordingly, these results indicate that certain anthropogenic polychlorohydrocarbons are a more serious threat to wildlife than methyl mercury. The latter is also a naturally occurring biocide, but it has caused serious mass poisonings only in connection with anthropogenic leakages of mercury (see Chapter 7).

## 6.5 DDE AND EGGSHELL THINNING EFFECT

An explanation for the decline of certain predatory birds in areas of intense agriculture was found in the eggshell-thinning effect caused by high DDE levels.[13,14] The effect is characterized by an eggshell index (EI) defined as follows:[15]

$$EI = \frac{\text{Weight of the shell (mg)}}{\text{Breadth (mm)} \times \text{Length (mm)}}.$$

As an example, DDE contents in the membrane lipids of peregrine falcon eggs collected at different times and from a variety of areas correlate significantly with EI. A present linear regression run from data presented by Peakall and Kiff[14] gives the following output:

**DDE and Eggshell Thinning of Peregrine Falcons**

| | |
|---|---|
| Linear regression $Y = A + B * X$, where $Y$ = Eggshell Index and $X$ = DDE ppm in egg lipid | |
| Mean $Y$ (Eggshell Index) = 1.6525 | |
| Standard deviation of $Y$ = .2576794 | |
| Mean $X$ (DDE in Lipid) = 92 | |
| Standard deviation of $X$ = 61.9295 | |
| Intercept | $A$ = 1.965416 |
| Slope | $B$ = −3.401263E-03 |
| Correlation coefficient | $R$ = −.817444 |

Graphic presentation of the data is shown in Figure 6.5.1.

## 6.6. PCB IN SEDIMENTS

Sediments represent an ultimate sink for persistent anthropogenic pollutants. Measures of the pollutant load in certain areas can be based on sediment analyses. Some of the most persistent and best-known pollutants are a mixture of polychlorinated biphenyls (PCBs). PCB analyses in sediment in the Great Lakes of North America showed that all residues were at 0 to 3 cm surface segment in the Upper Lakes (Superior, Michigan, Huron), but a fraction of PCBs were found also at deeper layers

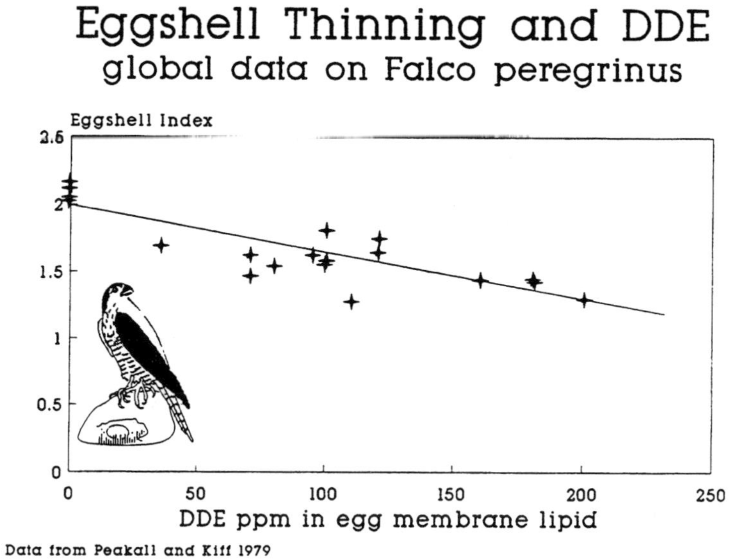


**Figure 6.5.1** Eggshell thinning and DDE levels. Description of the global data on *Falco peregrinus* from Peakall and Kiff.[14]

in the Lower Lakes (Erie, Ontario), where the flux of sedimentation is higher. However, in 1981, the sedimented burden of PCB in different lakes was estimated from surface sediment (0 to 3 cm) concentration.[16] The sediment density, 2.5 g/cm$^3$, and the known bottom areas of the lakes were used in calculation. The results were as follows:

| Lake | Concentration ng/g dw | Burden $10^3$ kg |
|---|---|---|
| Superior | 3.3 ± 5.7 | 20 |
| Michigan | 9.7 ± 15.7 | 42 |
| Huron | 13 ± 10 | 58 |
| Erie | 95 ± 114 | 182 |
| Ontario | 57 ± 56 | 115 |
| TOTAL | | 417 |

These and other results indicate that PCBs are only partly atmospherically transported pollutants; the local leakages have a significant role in their environmental occurrence. A very clear point-source-related case was found by sediment analyses in Lake Jyväsjärvi in Finland.[17] The sampling sites and results of PCB concentrations in sediment layers (3 cm slices) are shown in Figure 6.6.1.

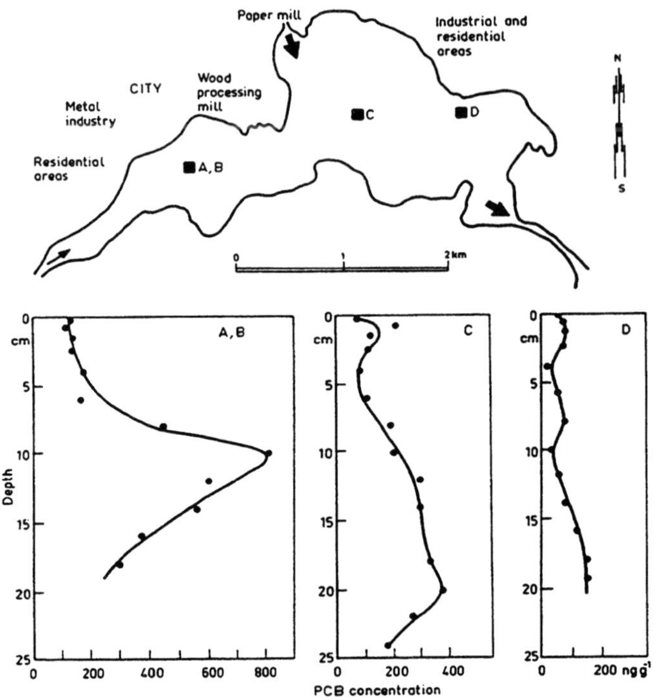


**Figure 6.6.1** Map of Lake Jyväsjärvi in the city of Jyväskylä showing the sediment sampling sites A to D and PCB concentration profiles (ng/g dw) in sediment cores in these sites. A and B represent two cores taken from the same site.[17]

The layers were dated by pollen stratigraphy. From dating data, the annual fallout of dry matter per square meter during the time period of each slice could be calculated. Multiplication of these fallout rates by concentration (dry weight basis) in the corresponding layer gives the annual fallout (sedimentation rates) of the pollutant measured (in this case, PCB). The rates obtained are presented in Figure 6.6.2.

There had been a local PCB release in the western part of the lake with a maximum in 1969 and a significant decrease thereafter. The PCB pollution in the eastern part of the lake (site D) after 1980 represents the atmospheric fallout, which was about 0.2 $mg/m^2$ per annum. Even this was twice as high as the general background fallout of PCB in the latitudes of central Finland. Remarkably, mother's milk in Jyväskylä, measured in 1984, contained 2.5 times the Finnish average level of PCBs.

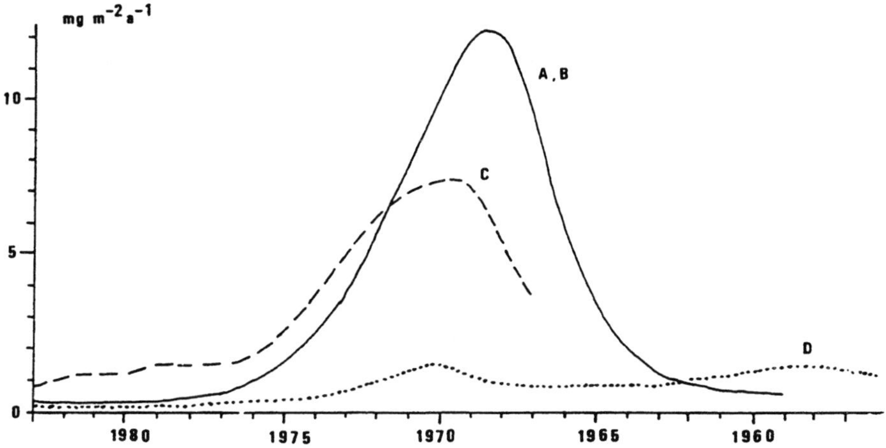


**Figure 6.6.2** Sedimentation rates (fallouts) of PCBs per square meter per annum in different years in three sampling sites (AB, C, and D of Lake Jyväsjärvi.[17]

## 6.7 SUMMARY

It is advantageous to consider the use of statistical methods in the environmental study plan before sampling. Concentration data to be handled are most often lognormally distributed in wildlife (and human) populations analyzed. Covariance analyses together with Q-tests can be used to estimate differences between populations. Various linear regressions are suitable for the estimation of effects and trends. Biomagnification can be estimated by comparing levels or, more reliably, amounts of compounds in specimens.

## 6.8 REFERENCES

1. **Snedecor, G. W., and W. G. Cochran.** *Statistical Methods*, 7th ed. (Ames, IA:Iowa State University Press, 1987).
2. **Breen, J. J. and P. E. Robinson, Ed.** "Environmental Applications of Chemometrics," *ACS Symp. Ser.* 292 (1985).
3. **Sielken, R. L., Jr.** "Statistical Evaluations Reflecting the Skewness in the Distribution of TCDD Levels in Human Adipose Tissue," *Chemosphere* 16:2135-2140 (1986).

4. **Paasivirta, J., and R. Paukku.** "Use of Composited Samples to Optimize the Monitoring of Environmental Toxins," *Chemosphere* 19:1551-1562 (1989).
5. *Chemical Trends in Wildlife. An International Cooperative Study* (Paris:OECD, 1980).
6. **Cooke, D., A. H. Craven, and G. M. Clarke.** *Basic Statistical Computing* (London:Edward Arnold Ldt, 1982).
7. **Herve, S., P. Heinonen, R. Paukku, M. Knuutila, J. Koistinen, and J. Paasivirta.** "Mussel Incubation Method for Monitoring Organochlorine Pollutants in Watercourses. Four-year application in Finland," *Chemosphere* 17:1945-1961 (1988).
8. **Paasivirta, J., S. Herve, P. Heinonen, and T. Rantio.** "Bioaccumulating Organochlorine Compounds in Finnish Watercourses Surveyed by the Mussel Incubations," Nordic Symposium on Organic Environmental Chemicals 11-14.9.1989, ABSTRACTS, University of Joensuu. Faculty of Mathematics and Natural Sciences Report 29, 65-66 (1989).
9. **Paasivirta, J., K. Heinola, T. Humppi, A. Karjalainen, J. Knuutinen, K. Mäntykoski, R. Paukku, T. Piilola, K. Surma-Aho, J. Tarhanen, L. Welling, H. Vihonen, and J. Särkkä.** "Polychlorinated Phenols, Guaiacols and Catechols in Environment," *Chemosphere* 14:469-491 (1985).
10. **Paasivirta, J., P. Klein, M. Knuutila, J. Knuutinen, M. Lahtiperä, R. Paukku, A. Veijanen, L. Welling, M. Vuorinen, and P. J. Vuorinen.** "Chlorinated Anisoles and Veratroles in Fish. Model Compounds. Instrumental and Sensory Seterminations," *Chemosphere* 16:1231-1241 (1987).
11. **Paasivirta, J., J. Särkkä, J. Pellinen, and T. Humppi.** "Biocides in Eggs of Aquatic Birds. Completion of a Food Chain Enrichment Study for DDT, PCB and Hg," *Chemosphere* 10:787-794 (1981).
12. Ministry of Interior, Finland. *Publication of the Environmental Protection Agency A:3* (1985) (in Finnish), and nonpublished results of J.Paasivirta et al.
13. **Ratcliffe, D. A.** "Decrease in Eggshell Weight in Certain Birds of Prey," *Nature* 215:208-210 (1967).
14. **Peakall, D. B., and L. F. Kiff.** "Eggshell Thinning and DDE Residue Levels Among Peregrine Falcons (*Falco peregrinus*): A Global Perspective," *Ibis* 121:200-204 (1979).

15. **Ratcliffe, D. A.** "Changes Attributable to Pesticides in Egg Breakage Frequency and Egg Shell Thickness in Some Birds," *J.Appl.Ecol.* 7:67-116 (1970).
16. **Eisenreich, S. J., and T. C. Johnson.** "PCBs in the Great Lakes: Sources, Sinks, Burdens," in *PCBs: Human and Environmental Hazards*, F. M. D'ltri and M. A. Kamrin, Eds. (Boston: Butterworth, 1983) pp. 49-75.
17. **Paasivirta, J., K. Mäntykoski, R. Paukku, T. Piilola, H. Vihonen, J. Särkkä, and K. Granberg.** "PCB in the Sediments of the Lake Jyväsjärvi," *Aqua Fennica* 16:17-23 (1986).

# 7

# Mercury in the Environment

## 7.1 INTRODUCTION

Mercury is the only metallic element that occurs as a liquid at room temperature. It is also highly volatile. Mercury was known by ancient Chinese, Near East, and Egyptian cultures as a metal and its principal ore cinnabar as a medicine and a source of red ink. Greeks in the fifth century used it as a pigment, and Aristotle described the metal. Later, mercury served in the metallurgy of gold and silver and as a central element in alchemy. The industrial revolution of the 18th century produced a significant increase of mercury emissions to the atmosphere by the use of fossil fuels. Other significant innovations that created environmental mercury problems were the production of chlorine and caustic soda by using mercury cathodes (1982) and the use of organomercury compounds as seed fungicides (1900) and as slimicides (1950). Anthropogenic and also naturally formed methyl mercury is a toxic and persistent chemical that is bioaccumulated and enriched in some food chains. Its hazardous and natural baseline levels are extraordinarily close to each other: their ratio is only 10 to 100.[1,2]

## 7.2 THE NATURAL MERCURY CYCLE

The main routes of the global cycling of mercury were shown in Figure 2.6.2 (Chapter 2). Most of the mercury on the earth appears as minerals in rocks. The average content is only 0.08 ppm, but mercury is enriched in volcanic sulfides up to percentage levels. Natural weathering of rock releases about 800 tons of mercury each year. The majority of this mercury reaches the sea largely in suspended solids. A small portion reaches the ocean in solution. A portion is deposited in sediments, and another portion is incorporated into soil. Uptake by plants is weak, but animals concentrate methylated mercury.

Natural enrichments in rocks in some areas cause concentrations of mercury in water and soils that are 100 times higher than average. However, these emissions have had little or no observable deleterious effects on their surroundings.[3]

About 16 billion tons of mercury has been released by weathering from the rocks during the geological history of earth. Only 0.03% of this amount, however, is found in solution in the oceans.[3] While volcanism and other natural phenomena have also released significant amounts of mercury, sedimentation has removed more than 99.97% of this dissolved mercury from the hydrosphere.

## 7.3 ANTHROPOGENIC INFLUENCE TO THE MERCURY CYCLE

### 7.3.1 Total Releases

Human activities increase the release of mercury to air, water, and biota (Figure 7.3.1). Between 1900 and 1970, natural weathering released 57,000 tons, but mining released 361,000 and fossil fuel combustion released 151,000 tons of mercury from the rocks to the global cycle.[3]

### 7.3.2 Chlorine Production and the Forest Industry

Most chlorine gas has been manufactured together with caustic soda by the electrolysis of sodium chloride using mercury cells as a cathode. In the 1960s, this caused the discharge of approximately 100 g of Hg per ton of chlorine produced. Most of this mercury consisted of effluent to watercourses, 5 to 10 g to chlorine gas, 0 to 10 g to caustic soda, and 15 to 25 g to air.[1] Contamination of chlorine and caustic soda was one reason for mercury spills from pulp mills into the receiving watercourses. Later, process improvements decreased these leakages to 0.5 g of Hg per ton of produced chlorine gas. This reduced the mercury effluents of chlorine and pulp mills to one hundredth of those of the 1960s.

Another forest-industry effluent problem arose from the use of organic mercury slimicides in paper production. This release was stopped in 1968 in Finland and in 1971 in the U.S.A.

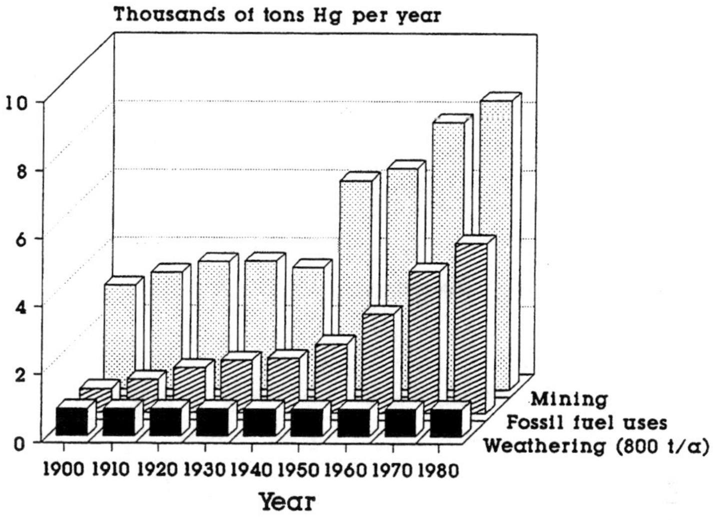


**Figure 7.3.1** Annual releases of mercury from the rocks.

## 7.3.3 Mining and Refining

Mercury spills from mining and ore refining are very significant. Stack losses are usually 2 to 3% of the total mercury handled. The atmosphere at ground level near mercury ore deposits can contain 16 ppb (weight basis) of Hg, while the background level is about 0.0001 ppb or less. In the neighborhood of mercury-ore mines and metallurgy plants, as well as near to natural point sources (e.g., volcanic emanations), increased levels up to 200 ppb of mercury in ground water have been measured. Mercury amalgams used in gold and silver metallurgy and in jewelry casting often cause high levels of mercury exposure to personnel and the environment.

## 7.3.4 Mercury Bactericides and Fungicides

The use of organomercurials as bactericides and fungicides in the paint industry and in agriculture contributes significantly to the mercury pollution of the environment. In the receiving waters of paint-industry plants, in the vicinities of shipyards, and in shallow harbors, increased mercury levels have been observed in fish due to the release of antifouling paints into the water.

Organomercuric seed dressing can introduce high levels of mercury into the soil. Seed-eating birds and their predators were shown to collect ecologically damaging amounts of methyl mercury in Sweden, where mercurial fungicides were therefore banned in 1966.

### 7.3.5 Laboratories, Hospitals, and Dentistry

Mercury is an important element in various types of laboratory equipment, drugs, reagents, dental preparations, disinfectants, and sterilizing solutions. These uses cause discharges of mercury into municipal waste-treatment plants. Raw sewage in the U.S.A. contains an average of 2 ppb of Hg, of which 1 ppb comes from the effluent of dental clinics, laboratories, and hospitals. Amalgam-originated Hg vapors have caused mercury poisoning symptoms in some dentists.

### 7.3.6 Combustion

The levels of mercury are 0.01 to 3.00 ppm in coal, several ppms in fuel oils, 1 to 20 ppm in gasoline, 0.03 to 0.5 ppm in raw materials for cement production, and some ppms in municipal refuse combusted in incineration. Most of this mercury is released into the atmosphere by energy production and other combustion processes. Anthropogenic combustions alone are, at present, a much more important source of mercury flow to the atmosphere and waters than all of natural weathering.

## 7.4. COMPOUNDS, REACTIONS, AND AMOUNTS OF MERCURY IN THE ENVIRONMENT

### 7.4.1 Mercury in the Atmosphere

Normally, atmospheric mercury consists of 49% free element, 25% mercury(II)halide, 21% monomethyl mercury, and 1% dimethyl mercury vapor, and 4% of particle-bound mercury.[1] Elemental mercury and its compounds are transformed into each other by chemical and photochemical reactions.

Mercury in air is in dynamic movement between water and soil. Its residence time in the atmosphere is 5.5 to 90 days.

Emissions from land degassing (17,800 t/y), volcanoes (20 t/y) and anthropogenic sources (10,000 t/y) contribute to this equilibrium.

The atmospheric mercury of Europe (outside the USSR) in the late 1970s was estimated in metric tons per year[4] to be:

| | | |
|---|---|---|
| Anthropogenic emissions | 500 | |
| Background emissions | 50 | > Total emissions 550 t/y |
| Wet deposition | 100 | |
| Dry deposition | 100 | > Total deposition 200 t/y |

Preindustrial deposition has been estimated to be 50 to 100 t/y or 10 to 20 μg/m². The content of mercury in air on the continent decreases with altitude according to the relationship

$$C_z = C_0 e^{-kz}$$

where $C_0$ is the ground concentration and $C_z$ is the concentration at an altitude $z$; $k$ is a constant of the order 1 mm.[2] Ground and air temperatures significantly affect concentrations, which are higher during warm weather periods.

The background air concentration of Hg varies on the Atlantic Ocean from 0.4 to 20 ng/m³. High Hg levels in air have been measured in densely populated areas (150 to 550 ng/m³), above fields where mercury fungicides are used (10,000 ng/m³), on mines up to 1 million ng/m³, and in volcanic gases 100 to 40,000 ng/m³.

### 7.4.2 Mercury in the Hydrosphere

Mercury enters the water compartment by deposition (rainwater contains 0.0008 to 1.45 μg/l Hg) and as silted or dissolved emissions from natural weathering and anthropogenic spills. Mercury content in water varies greatly depending on the degree of pollution. Measured content in snow, surface freshwaters, groundwaters, and seawater ranges from 0.08 to 1.1, 0.01 to 8.4, 0.01 to 0.46, 0.001 to 1.6 μg/l, respectively.[1-3]

Ratios of Hg compounds in water depend on the oxygen content, pH, and biological activity of the watercourse. Reduced sulfur has a high affinity to mercury: both inorganic and organic SH compounds remove mercury from solution. $Hg^{2+}$ binds SH most strongly. The high toxicity of $Hg^{2+}$ salt is explained by its inhibitory effect on enzyme proteins containing SH groups.

Another path of removal of mercury from water is its microbial methylation followed by bioaccumulation of the lipophilic methyl mercury formed. Organic mercury composes up to half of the total mercury in river and coastal sea waters.[1] However, in biota, the majority of the mercury is organic.[2,3,5,6]

In the surface layer of the oceans, the mercury content is about 0.1, but at greater depths it increases to 0.15 to 0.27 μg/l due to the uptake of mercury by plankton at the surface and transport to deeper regions via plankton sinking.[1]

### 7.4.3 Mercury in the Sediment/Soil Environment

Sediments collect most of the mercury moving in the hydrosphere. Background levels of Hg in clean lake sediments are 0.05 and in oceans are 0.1 to 1 μg/g (dry weight). Anthropogenic releases have increased concentrations significantly. The mercury content in recent sediment layers in Lake Ontario, in lakes of Switzerland, and in the River Rhein near Koblenz is 0.31 to 1, 0.01 to 2.23, and 4.5 μg/g, respectively. In sediment of the Wisconsin River, mercury concentration is generally 0.4 to 2.7 μg/g, but near a chloroalkaline plant 684 μg/g.

Terrestrial soil strongly adsorbs the deposited, agriculturally applied, and waste-originated mercury. Surface soils are at present richer in mercury than deeper layers. This can be explained by the strong adsorption and slow desorption of mercury, a process that also involves ion exchange phenomena.[2] Consequently, increased pollution leads to increased mercury deposition, which accumulates in surface layers due to extremely slow leaching of mercury in the soil column.

### 7.4.4 Mercury in the Biosphere

Plankton are the first collectors of mercury on its way to the biosphere. Bioconcentration factors (Bcf = concentration in plankton/concentration in water) as high as 100,000 have been measured. The influence of transport from the terrestrial environment of industrialized areas is very significant. This has

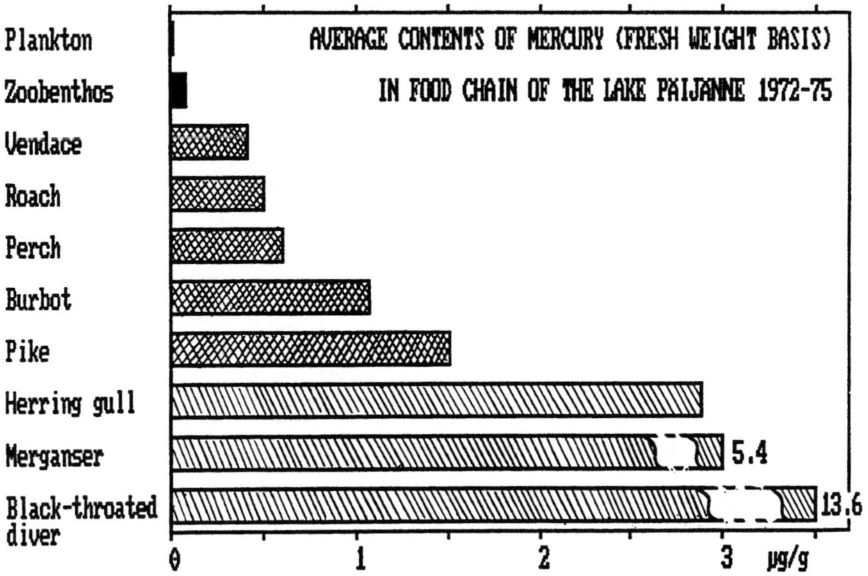


**Figure 7.4.1** Overall food-chain enrichment of mercury as the content in species of different trophic levels in a slightly mercury-polluted lake ecosystem.[6]

been demonstrated in Atlantic plankton, where the mercury content decreases exponentially with increasing distance from the east coast of the U.S.A.[1]

Hg in fish is almost completely organic. Lipophilicity of methyl mercury leads to the high Bcf. Bcf values of fish for the mercury uptake through gills are 5,000 to 100,000. But mercury enriches by food-chain transfer as well. Algae-eating fish generally have 0.001 to 0.2 and their predators (pike, tuna, shark) 0.01 to 6 μg/g Hg in fresh muscles. Below point sources such as chloralkali plants, a content of more than 10 μg/g has been measured.

Mercury content further increases from fish to fish-eating birds (Figure 7.4.1). The nature of the biomagnification of mercury is clearly seen in this figure: the content in plankton feeders (vendace) is lower than in small predators (perch), in the zoobenthos predator (burbot), and in the large predator (pike). Fish-eating birds (merganser and black-throated diver) collect more mercury than the herring gull, which also eats terrestrial species and garbage.

Factors causing the food-chain enrichment are the biological

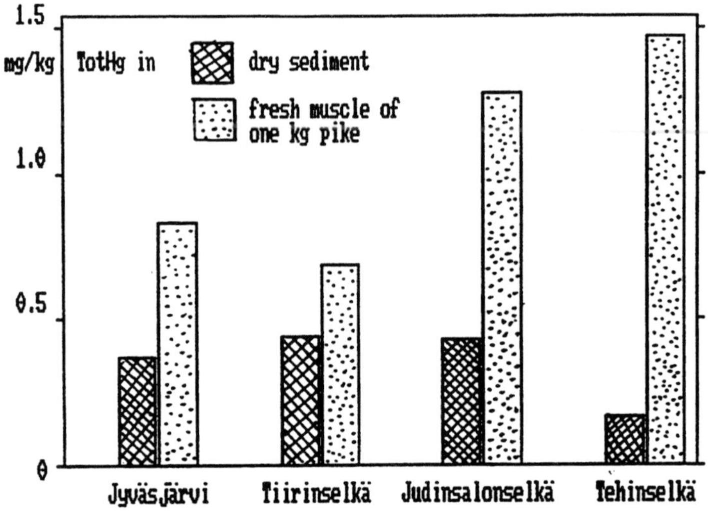


**Figure 7.4.2** Comparison of average mercury levels in surface sediment and pike at four lake areas from 1972 to 1974. Jyväsjärvi and Tiirinselkä were polluted by urban and industrial wastes. Judinsalonselkä was only slightly polluted, and Tehinselkä was in nearly natural condition.[5-7]

persistence and slow excretion of methyl mercury in organisms. Its lifetime measured in fish is 400 to 1000 days but in humans only 70 to 76 days.

Mercury contamination in the sediment of Lake Päijänne was also studied in 1972 to 1974.[7] An example of the results compared to the mercury levels in pike at four selected areas is illustrated in Figure 7.4.2.

The high mercury content of the surface sediment did not correspond to the high content in fish in the same lake area. The lake areas in question had been under heavy mercury pollution until 1968 from slimicide uses of the paper industry. After these spills stopped, mercury levels in fish were regularly decreasing.[8] However, the lowest levels appeared in pike of Tiiriselkä, where the sedimentation of sulfur-containing particles was highest. These particles obviously assisted removal of mercury from water solution by sorption and chemical binding to reduced sulfur.

Another factor in the transfer of mercury to biota is desorption of (methyl) mercury from soil and recent sediment. Lakes having high contents of humus in the water and newly im-

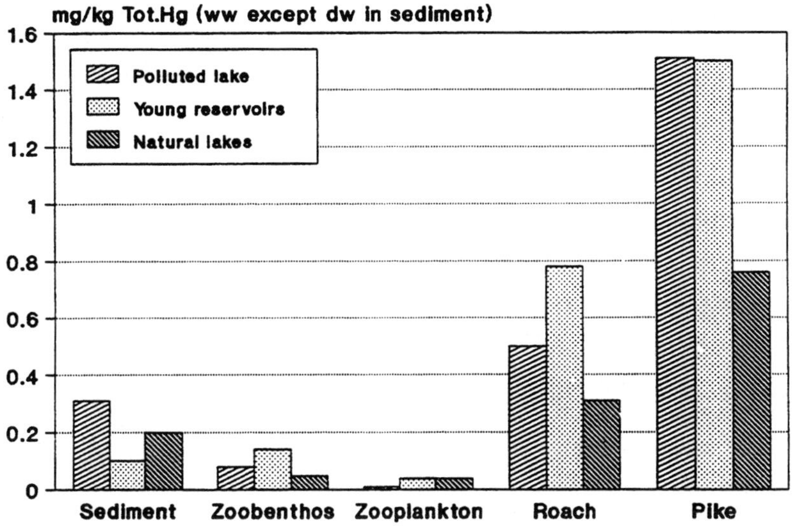


**Figure 7.4.3** Average total mercury levels in biota of a polluted lake (Päijänne), new (young) reservoirs, and natural lakes from 1980 to 1985 in Finland.[9]

pounded reservoirs have elevated levels of mercury in their aquatic fauna (Figure 7.4.3).

The fraction of the mercury present as organic mercury is higher in biota than in soil or sediment. Methyl mercury is nearly insoluble in water, but it forms colloids with humic material. This leads to an increased concentration of mercury in the water phase. Humic substances transfer mercury from disturbed soil to water and consequently into the food chain.[9,10]

The enrichment of Hg from fish to birds is demonstrated by comparing the total mercury amounts in eggs, juveniles, and adults at different lakes (Figure 7.4.4).

## 7.5 ANALYTICAL METHODS FOR MERCURY

### 7.5.1 Determination of Total Mercury[1,2]

For analysis of all mercury compounds together, an environmental sample is treated most commonly by oxidants in liquid phase. In this "wet oxidation," many reactants are available. $H_2F_2$, HCl, $HClO_3$, $HClO_4$, HBr, $HBrO_3$, $H_2SO_4$, $H_2O_2$, $KMnO_4$, $K_2Cr_2O_7$, $K_2S_2O_8$, $V_2O_5$, and their mixtures have been used. Open systems frequently used in wet combustion involve the risk of outside contamination and volatilization of mercury. There-

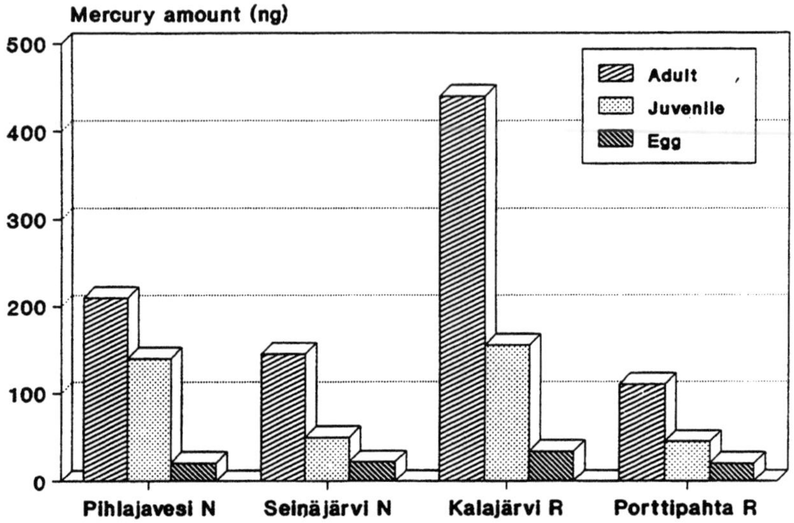


**Figure 7.4.4** Total mercury amounts in eggs, juveniles, and adults of goldeneyes at two natural lakes (N) and at two reservoirs (R). Pihlajavesi was more humic than Seinäjärvi. Kalajärvi Reservoir was one to three years old, Porttipahta Reservoir was four to six years old at the time of the study.[9]

fore, vessels with reflux or pressure decompositions in completely closed devices are preferred. Organomercurials dissolved in water can be degraded, e.g., with ultraviolet irradiation.

Separation of mercury from the matrix is most often performed by volatilization with heating at 900°C in nitrogen, air, or argon stream. Hg is separated from organic solids and coal by combustion in high-frequency- or ultra-high-frequency-induced oxygen plasma. From aqueous solutions and solutions made in decomposition, ionic mercury is reduced to elemental mercury and removed by aeration (cold vapor technique). Other separation methods used are solvent extraction of organomercurials, precipitation as HgS from aqueous solution, liquid-liquid extraction with dithizone or other complex forming agents, by ion exchange, thin layer, paper, or gas chromatography, by electrophoresis or by electrodeposition, as ultramicro method in the lower ng/g and pg/g range.

Gaseous mercury can be preconcentrated by passing the gas stream through impringer flasks containing permanganate sulfuric acid, iodine potassium iodine, iodine monochlorine, etc., as absorbents. Solid adsorbers such as Cu, Ag, Au, Pt, Chromo-

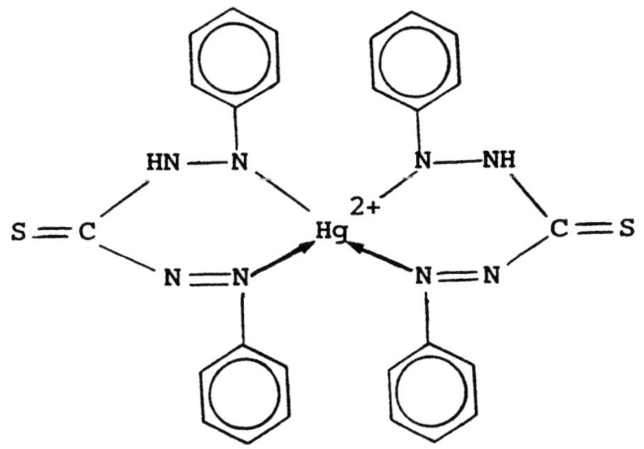


**Figure 7.5.1** Complex of dithizone with $Hg^{2+}$.

sorb W, activated charcoal with KJ or CdS, Au-impregnated filters, glass fibers, glass wool, or glass beds separate mercury from combustion gases and allow the preconcentration.

Mercury vapors can be trapped in a cooled capillary and the diameter of the metal droplets measured under a microscope. In this way, the mercury can be determined in the μg/g range.

The most common determinations of mercury are done by titrimetric, colorimetric, and spectrometric methods. Dithizone or diphenyl thiocarbazone is very widely used in both isolation and in titrimetric or colorimetric measurement of mercury. The complex A formed by dithizone with $Hg^{2+}$ (Figure 7.5.1) absorbs light with a maximum at a wavelength of 490 nm, and its $CCl_4$ and $CHCl_3$ solutions are orange yellow.

The methyl mercury complex of dithizone has an absorption maximum at 476 nm. Another mercury complex used in colorimetry is formed from mercury silver iodide and crystal violet. It has a light absorption maximum at 605 nm.

The most popular method for final determination of mercury is the atomic absorption spectrometry (AAS). In AAS, the metal to be measured is atomized in flame or in a graphite oven. The absorption of electromagnetic radiation by the atomic vapor is then measured by the spectrometric detector, e.g., a photomultiplier. The wavelength used for AAS of mercury is normally 253.7 nm. However, more sensitivity is achieved at 184.9 nm, where a purge with $N_2$ or with a noble gas is necessary. AAS is often closely connected with the separation and preconcentration steps in devices performing multistage procedures for mercury analysis.

Atomic emission (AE) from plasmas can be used for Hg measurement. Microwave-induced gas plasmas (MIP), mainly of helium or argon produce the best sensitivity (pg/g). Somewhat poorer sensitivity is obtained with inductively coupled plasmas (ICPs), radio-frequency-coupled plasmas (RFCPs), or arc-excited plasmas.

Atomic fluorescence (AFS) and X-ray fluorescence methods are fairly sensitive — nanogram per gram levels of mercury can be detected.

Electroanalytical methods, potentiometry, conductometry, DC- and AC-polarography, amperometry, and anodic stripping methods allow determination of mercury in the μg/g to ng/g range.

Neutron activation analysis (NAA) and catalytic methods (mercury catalysis or inhibition in certain reactions) are also used for determinations of mercury in ng amounts.

### 7.5.2 Analysis of Individual Mercury Compounds

Environmental mercury occurs as elemental vapor, inorganic salts, and organic compounds. Specific analyses are necessary to monitor the risk of mercury contamination. From air, the individual forms can be adsorbed or absorbed specifically. For liquids, TLC, paper chromatography, and electrophoresis are used with different coatings and eluents and also with complex formation to separate different mercury compounds for determination.

Gas chromatography (GC) with a flame-ionization detector (FID) or with an electron-capture detector (ECD) is an efficient and sensitive method to measure mercury compounds in complex mixtures. Other useful methods are high-performance liquid chromatography (HPLC) with voltammetry and GC with flameless AAS, AE-MIP, or mass spectrometry. These methods have determination limits as low as pg/g in some cases.

In environmental or human health monitoring, the very necessary determination of inorganic and organic mercury from a single sample can be done with high sensitivity by a simple procedure connected to cold-flame AAS.[9,11] In this method, the wet sample is first degradated by NaOH digestion, in which organic mercury remains unchanged. Then inorganic

**Table 7.6.1 Trends of Hg in marine (M), freshwater (Fw) and terrestrial (T) locations studied in 1972–75 by OECD[12]**

| Number of trend analyses | | | | | | | | | | | | |
|---|---|---|---|---|---|---|---|---|---|---|---|---|
| Increasing | | | Decreasing | | | No Trend | | | Total Analyses | | | |
| M | Fw | T | M | Fw | T | M | Fw | T | M | Fw | T | Total |
| 2 | 3 | 0 | 3 | 7 | 5 | 3 | 4 | 0 | 8 | 14 | 5 | 27 |

mercury is reduced to its elemental form by $SnCl_2$ and simultaneously evaporated and measured. Finally, organic mercury is reduced with $NaBH_4$ to Hg vapor which is measured.[11] Successful modifications of this method, e.g., the use of more handy reducing agents, have been developed.[9]

## 7.6 ANALYSES OF TRENDS

### 7.6.1 OECD Monitoring[12]

After measures to limit mercury effluents were implemented, monitoring studies were done to follow whether these measures have any influence on mercury levels in wildlife. A comprehensive study was the Wildlife Analysis and Sampling Programme 1972 to 1975 of the Organization for Economic Cooperation and Development (OECD).[12] Indicator species from the terrestrial ecosystem were juvenile starlings, from the marine ecosystem herring, and from the freshwater ecosystem perch. Each year 25 specimens from each location were analyzed. Duplicate determinations were done from five specimens to estimate the laboratory variance (LV). The results were handled by statistical methods. The method used to determine the significance of trend was the Stepwise Multiple Regression Analysis. The total number of localities studied for mercury was 27 (Table 7.6.1).

The test animals were selected especially due to their worldwide occurrence and for monitoring of organochlorine residues. Therefore, mercury trends in many locations appeared to be nonsignificant. For example, perch in Finnish freshwater environments gave no significant trend, not even when the study was extended into 1978.[13] In contrast, for Hg in starlings of South Finland, extended trend analysis from 1967 to 1983 indicated a significant (**; $p < 0.01$) decreasing trend.[14]

### 7.6.2 Monitoring Using Pikes

Monitoring of local water areas for chemical pollutants is preferably done by analyzing pike, large, long-lived predatory fish which do not move long distances from their home area. After mercury release was halted at Päijänne in 1968, the trend in mercury there was followed intensively by analyzing pikes at three subareas of this 100-km-long lake. All results during each year were incorporated into trend calculation. When the influence of weight of the fish was eliminated in covariance analysis, the time trend 1970 to 1980 was found to be decreasing in all three areas with significance of *** ($p<0.001$).[8]

This trend analysis method also produced a reliable prediction for the future mercury levels in Päijänne pikes, because no new leakages appeared or were expected. Regression analysis was performed to calculate the dependence of mercury content on weight and year in each area. The results were extended to future years. In 1980 the following prognosis equations were obtained:

| Area | Total Mercury Content |
|---|---|
| Tiirinselkä | $TOTHG = 0.1094 * WEIGHT^{0.32732} * (YEAR-1970)^{-0.24242}$ |
| Ristinselkä | $TOTHG = 0.05932 * WEIGHT^{0.47422} * (YEAR-1970)^{-0.33158}$ |
| Tehinselkä | $TOTHG = 0.1758 * WEIGHT^{0.40841} * (YEAR-1970)^{0.-0.52466}$ |

The prognoses serve as estimates for future mercury levels of pikes to be consumed as food. For fast estimation, multidimensional projections of the above equations can be presented, e.g., as in Figure 7.6.1. Validity of the prognosis was investigated in 1981 by collecting ten pike from Ristinselkä, analyzing Hg from their muscles, and calculating (by linear regression) the observed TOTHG for 1-kg pike. The predicted content was 0.70, and the observed was 0.69 mg/kg. Then the prognosis was revised adding the new data to previous ones. The following equation was obtained:[15]

$$\text{Ristinselkä } TOTHG = 0.0650 * WEIGHT^{0.460} * (YEAR - 1970)^{-0.331}$$

A second validation study was done in 1987 by analyzing again ten pike from Ristinselkä. Predicted TOTHG for 1-kg pike

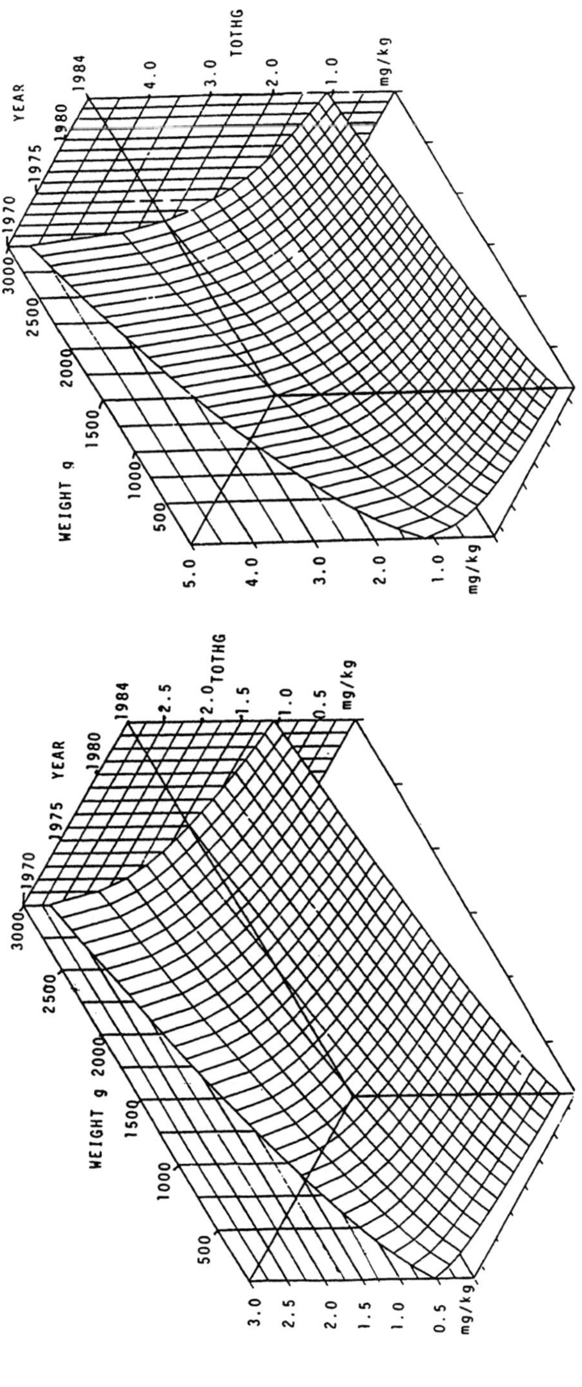


**Figure 7.6.1** Mercury content in pike estimated in 1980 as function of weight and time.[8]

then was 0.61 mg/kg. Linear regression of the analysis results gave as the observed TOTHG value for 1-kg pike 0.61 mg/kg!

This method is useful in planning and managing future fishery economy in new reservoirs, which normally are rich on fish but have mercury problems in the first years after impoundment.

## 7.7 ENVIRONMENTAL MERCURY AS A HUMAN HEALTH HAZARD

### 7.7.1 Methyl Mercury Poisoning

The greatest environmental mercury hazard to man is bioaccumulation of methyl mercury in the body by ingestion of contaminated food. Occupational exposures by skin adsorption and inhalation are other hazards.

The main symptoms of methyl mercury poisoning are loss of sense in tongue and lips, ataxia (loss of coordination in gait), visual field constriction, hearing loss, symptoms of a disturbed nervous system, and mental disturbances. Individuals who died of mercury poisoning had 1.4 or more μg/g Hg in their blood. Oral $LD_{50}$ of rat is 30 mg/kg. The common ADI (Acceptable Daily Intake) value given is about 0.3 mg/70 kg.[1,2,15]

### 7.7.2 Minamata Disease

A factory in Japan, which discharged mercury-containing waste into Minamata Bay, caused the disaster known as Minamata disease. The first case appeared in 1953. The epidemic came to the attention of public health authorities 1956, when the number of cases was 80. The symptoms were identified with those found in accidental alkyl mercury poisoning in 1958, and in next year the causative agent was shown to be methyl mercury present in the fish and shellfish of the bay. A similar catastrophe occurred in Niigata, Japan, in 1964 to 1965. The Minamata poisoning involved 121 cases with 46 deaths, and the Niigata poisoning 47 cases with 6 deaths.[2,16]

Poisonings of animals have been common. Cats in Minamata and in some places in the U.S.A. had fatal poisonings that appeared as unsteadiness, frequent falls, circling movements, and convulsions. In the U.S.A., a herd of 44 adult swine and 5 litters of pigs were placed for several weeks on a diet half of which was grain treated with methyl mercury. Most of them died in 3 months.[2]

Like humans, cats and birds are sensitive to mercury poisoning, but fish are generally unaffected by high mercury levels. This makes it necessary to control seafood for mercury. Authorities in many OECD countries (e.g., in Finland) have recommended that fish containing 0.5 to 1.0 mg/kg Hg (in fresh muscle) should be eaten only once in two weeks and fish containing more than 1.0 mg/kg Hg should not be consumed at all.

In many areas of the U.S.A., Canada, Sweden, and Finland, elevated levels of methyl mercury in fish up to several ppm have been measured. Before official limitations of Hg levels in food, consumers of fish were found to have increased mercury levels in their bodies. Concentrations in hair were as high as 10 ppm. However, observations of any mercury poisoning symptoms in these populations were not reported.

### 7.7.3 Human Catastrophes from Treated Grain

Although more accidental than environmental exposures, poisonings of humans who consumed organomercurial fungicide-treated grain are a clear warning of risk caused by methyl mercury in the environment.

The largest catastrophe took place during the winter of 1971 to 1972 in Iraq, where people ate homemade bread prepared from wheat seed that had been treated with methylmercurial fungicide. Ninety-five thousand metric tons of such grain had been sold. Altogether, 6530 individuals were hospitalized and 459 of them died as a result of central nervous system damage.[2]

In Mexico in 1969 a farmer and five of his neighbors fed treated grain to hogs and then ate the contaminated pork. The pigs died or developed blindness, and several family members suffered methyl-mercury poisoning.

## 7.8 SUMMARY

Mercury and its inorganic and organic compounds have a natural cycle that is seriously affected by anthropogenic activities. Methyl mercury forms an ecotoxic hazard due to its biomagnification. Limitations of mercury releases have led to decreasing trends of mercury levels in biota. Many accidents with mercury contamination have caused fatal poisoning in animals and humans.

## 7.9 REFERENCES


1. **Kaiser, G., and G.Tödl.** "Mercury," in *The Handbook of Environmental Compounds, Vol. 3, Part A: Anthropogenic Compounds,* O. Hutzinger, Ed. (Berlin:Springer, 1980) pp. 1-58.
2. **Mitra, S.** *Mercury in the Ecosystem* (Aedermannsdorf, Switzerland:Trans Tech Publications Ltd, 1986).
3. **Gavis, J., and J. F. Ferguson.** "The Cycling of Mercury Through the Environment," *Water Res.* 6 :989-1008 (1972).
4. **Semb, A., and J. M. Pacyna.** "Toxic Trace Elements and Chlorinated Hydrocarbons: Sources, Atmospheric Transport and Deposition." Nordic Council of Ministers, Raport NORD 1988:74.
5. **Hattula, M.-L., J. Särkkä, J. Janatuinen, J. Paasivirta, and A. Roos.** "Total Mercury and Methyl Mercury Contents in Fish from Lake Päijänne," *Environ. Pollut.* 17:19-29 (1978).
6. **Särkkä, J., M.-L. Hattula, J. Paasivirta, and J. Janatuinen.** "Mercury and Chlorinated Hydrocarbons in the Food Chain of Lake Päijänne, Finland," *Holarctic Ecol.* 1:326-332 (9178).
7. **Särkkä, J., M.-L. Hattula, J. Janatuinen, and J. Paasivirta.** "Mercury in Sediments of Lake Päijänne, Finland," *Bull. Environ. Contam. Toxicol.* 20:332-339 (1978).
8. **Paasivirta, J., J. Särkkä, M. Aho, K. Surma-Aho, J. Tarhanen, and A. Roos.** "Recent Trends of Biocides in Pikes of the Lake Päijänne," *Chemosphere* 10:405-414 (1981).
9. **Surma-Aho, K., J. Paasivirta, S. Rekolainen, and M. Verta.** "Organic and Inorganic Mercury in the Food Chain of Some Lakes and Reservoirs in Finland," *Chemosphere* 15:353-372 (1986).
10. "Finland. Reports of the Mercury Project. Publications," National Board of Waters, Publ. Water Res. Inst. 65:1-107 (1986).

11. **Oda, C. E. and J. D. Ingle, Jr.** "Speciation of Mercury by Cold Vapour Atomic Absorption Spectrometry with Selective Reduction," *Anal. Chem.* 53:2305-2309 (1981).
12. *Chemical Trends in Wildlife. An International Cooperative Study.* (Paris:OECD, 1980).
13. **Paasivirta, J., and R. Linko.** "Environmental Toxins in Finnish Wildlife. A Study on Time Trends of Residue Contents in Fish During 1973-1978," *Chemosphere* 9:643-661 (1980).
14. **Paasivirta, J., R. Paukku, K. Surma-Aho, and L. Welling.** "Chemical Trends in Finnish Wildlife: A Study on Time Trends in Starlings During 1967-1983," *Chemosphere* 14: 457-468 (1985).
15. **Paasivirta, J., J. Särkkä, K. Surma-Aho, T. Humppi, T. Kuokkanen, and M. Marttinen,** "Food Chain Enrichment of Organochlorine Compounds and Mercury in Clean and Polluted Lakes of Finland," *Chemosphere* 12:239-252 (1983).
16. *Studies on the Health Effects of Alkylmercury in Japan* (Environmental Agency, Japan. 1975).

# 8

# Organohalogen Compounds in the Environment

## 8.1 INTRODUCTION

Halogenated natural compounds are rare in the terrestrial environment, but marine organisms produce a wide variety of them.[1] Some natural bromine and chlorine compounds are potent biocides. For example, the potent fungal antibiotics griseofulvin and aureomycin contain a chlorophenyl group in their structure. However, all known significant environmental toxicants among organohalogen compounds (OHCs) are anthropogenic.

OHCs may enter the natural environment through accidents, but more commonly they do so from industrial discharges, pesticide uses, urban wastes, and especially chlorination and combustion processes, and also from the forest industry. Observed long-term adverse effects on human health and nature are often associated with food-chain accumulation of polyhalogen hydrocarbons used as pesticides or in technical equipment, as, for example, polychlorinated biphenyls (PCBs) as heat transfer or capacitor fluids. Also many other polyhalogenated aromatic compounds have been found to be very toxic, persistent, and bioaccumulating.[2]

## 8.2 ECOTOXIC HALOGEN HYDROCARBONS

The harmful ecological effects of organochlorine compounds were first shown by the declines of certain bird populations in areas where organochlorine pesticides were heavily used, as Rachel Carson revealed in her book *Silent Spring* in 1962. A scientific explanation for this decrease in reproduction was found in the eggshell thinning effect of DDE, which is the major persistent metabolite of the insecticide DDT (see Sections 4.2 and 6.4). High levels of DDE and other persistent organo-

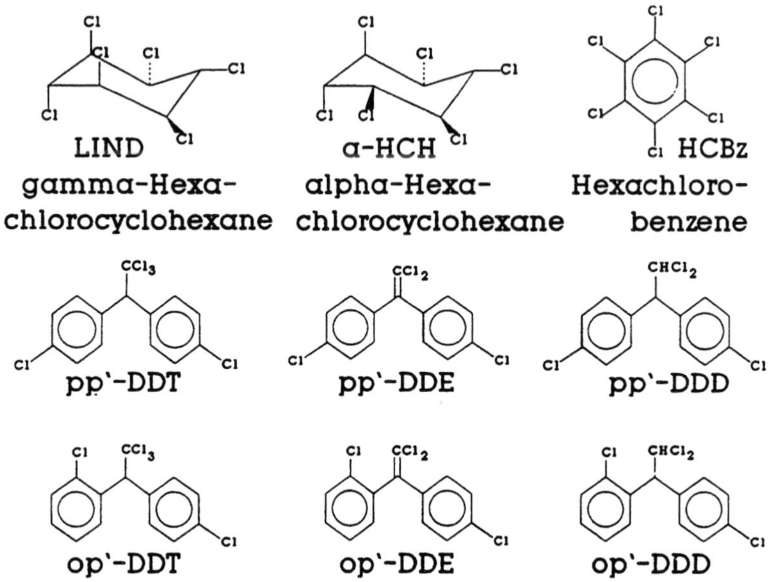


**Figure 8.2.1** Formula and name abbreviations of the persistent residues of organochloride pesticides lindane, hexachlorobenzene, and DDT.

chlorine pollutants threatened white-tailed eagles in the Baltic sea environment in 1960 to 1970. The population was saved due to the restriction of DDT use and successful feeding of eagles with nonpolluted meat.[2]

In addition to DDT residues, other polychlorinated insecticides and technical organohalogen hydrocarbons have been accumulated in food chains globally. Some of the most persistent or toxic OHCs are shown in Figures 8.2.1, 8.2.2, and 8.2.3.

o,p′-Substituted DDT traces in the environment originate from impurities in technical DDT preparations (p,p′-DDT as main active ingredient) used as insecticides.

After the restriction of DDT use in 1970, other polychlorinated hydrocarbons and their derivatives have been used in increasing amounts as pesticides. Of them, lindane has had the largest usage.

1,2,3,4,5,6-Hexachlorocyclohexane (HCH) has eight possible structural isomers. Only gamma-isomer (lindane) is active as insecticide. More than a million tons of technical HCH preparations containing 13 to 99% of lindane have been used globally from 1942 to 1980.[3] Being rather volatile and nonpolar, lindane occurs as an important organochlorine trace in the atmosphere

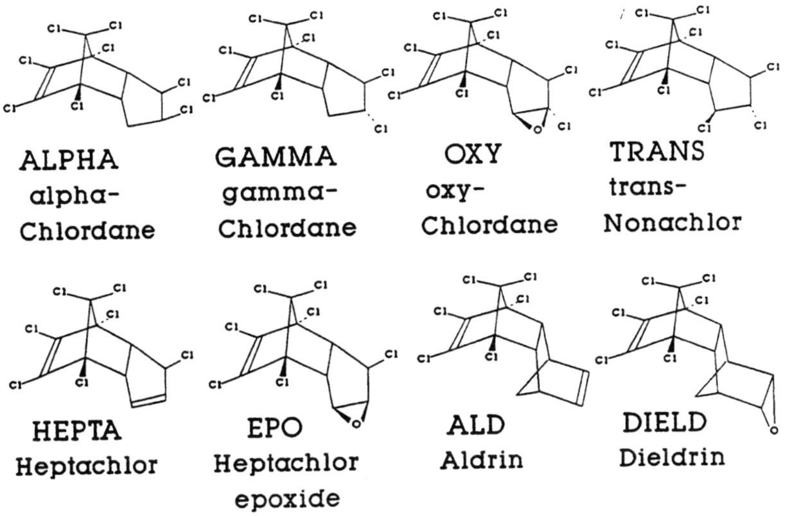


**Figure 8.2.2** Formula and name abbreviations of some cyclodiene insecticides. Sum of chlordanes SCHL = ALPHA + GAMMA + OXY + TRANS + HEPTA + EPO.

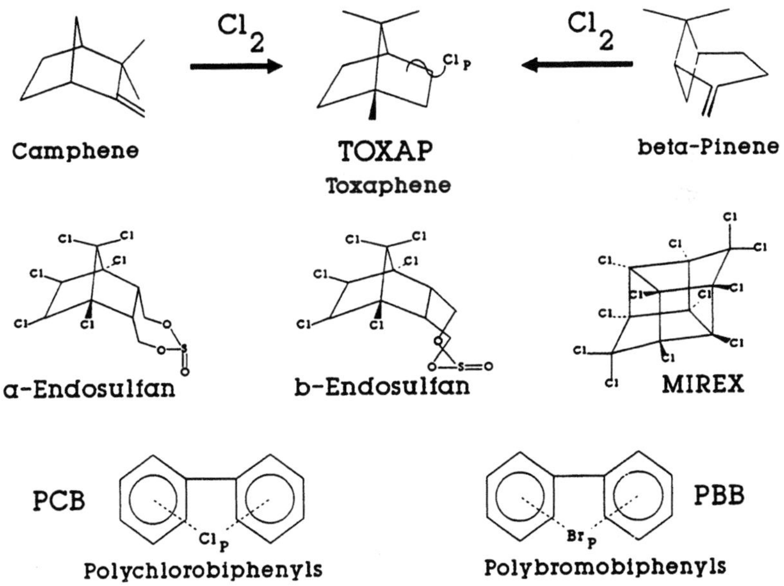


**Figure 8.2.3** Formula and name abbreviations of toxaphene, its terpene raw materials, endosulfan isomers, Mirex, PCBs, and PBBs.

(see Figure 3.5.1). Its stable isomer alpha-HCH is equally abundant.

Hexachlorobenzene (HCBz) occurs in the environment in amounts that are orders of magnitude higher than its production for fungicidal uses and technical fluids. Therefore, formation of HCBz as an unwanted by-product must be the major source of this occurrence.[4] It has been observed that large amounts of HCBz are formed by fast high-temperature combustion of chlorine-containing materials.[5]

Cyclodiene insecticide residues (Figure 8.2.2) are another group of global organochlorine pollutants. Chlordanes were never used in Scandinavia (Norway, Sweden, and Finland), but their residues are found there in rainwater and in biota.[2] An efficient cotton weed killer, toxaphene (Figure 8.2.3), is a complex mixture of polychloroterpenes that is prepared by chlorination of camphene or β-pinene. It replaced DDT in the U.S.A. and has been used increasingly in Central and Eastern Europe since 1970.

Endosulfan (Figure 8.2.3) is an insecticide/acaricide that is extremely toxic to fish. Mirex is a pesticide with a wide application range. It is only moderately toxic, but very persistent.[6]

PCBs as environmental contaminants were first found in Baltic seals and herring.[7] About a million tons of PCBs were produced from 1929 to 1987, of which 60% was used in electrical equipment, 15% in closed power and heat transfer systems, and 25% in many dispersive uses (plasticizers, paint and copy paper additives, adhesives, etc.).[3] PCB components having five or more chlorines (P = 5 to 10) are extremely persistent and are enriched in biota. While most PCB congeners are relatively nontoxic, some of them are potent dioxin-like toxicants (see Section 4.6).

Polybrominated biphenyls (PBBs) were the fire-retardant compounds in the Firemaster formulation that was accidentally mixed with cattle food in Michigan 1972. A mass poisoning of animals and long-lasting contamination of 9 million people by PBBs followed this episode. Many other areas of the U.S.A. have also been contaminated by PBBs.[8]

Polychloronaphthalenes (PCNs) and -terphenyls (PCT; three successive benzene rings) have been produced, both about 5000 t annually, for similar technical uses as PCB. Polychloroparaffins (PCPFs) are produced in amounts of 230,000 t/a to oil additives and plasticizers. PCN, PCT, and PCPF traces are commonly found in biota.[3,9]

## 8.3 FOOD CHAIN ENRICHMENT AND ECOTOXIC EFFECTS

Enrichment of polyhalogenated hydrocarbons is a very common phenomenon, although differences between specific compounds and species occur. Since 1960, many ecotoxic effects associated to high levels of these persistent traces have been observed.

Different DDT residues are enriched in the food chain at different rates depending on their persistency. Consequently, the most persistent metabolite DDE is the main residue in species at the highest trophic levels. Its high level causes the eggshell thinning effect in certain bird species (see Section 6.4).

Chlordanes and PCBs have equal or higher trophic enrichment in comparison to DDE (see Section 6.4). This was supported by analyses of Baltic birds and their eggs (Figure 8.3.1).[10]

High PCB levels have been associated with cases of reproduction impairment of Baltic seals and mink in Sweden, sea lions in California, and birds in Lower Great Lakes in North America. However, further studies have indicated that loss of the embryos and teratogenic effects might have been caused by impurities such as PCDDs and PCDFs, which often are found in PCB mixtures.[2]

Some symptoms of the heavily PCB-contaminated herring gulls of Lake Ontario and Lake Erie were identical to those of chickens in the U.S. broiler industry incident in 1957, which caused the deaths of millions of chickens.[11] The "chick edema disease" was a consequence of the leakage of heated PCB in broiler food production from the heat transfer device of cooking vessels. In 1970, it was shown to be caused by PCDFs formed in pyrolysis of PCB with oxygen. Toxic PCDDs caused identical symptoms. Therefore, embryonic mortality and tissue damages of the gulls could be associated with dioxin toxicity (see Section 4.6) of the persistent polyhalogenated traces enriched in birds to harmful levels. 2,3,7,8-TCDD, in fact, was analyzed in gull eggs at effective levels (1 ng/g) in 1972. Later, when TCDD levels decreased, the reproduction in gull colonies improved and the chick edema disappeared in 1975.[11]

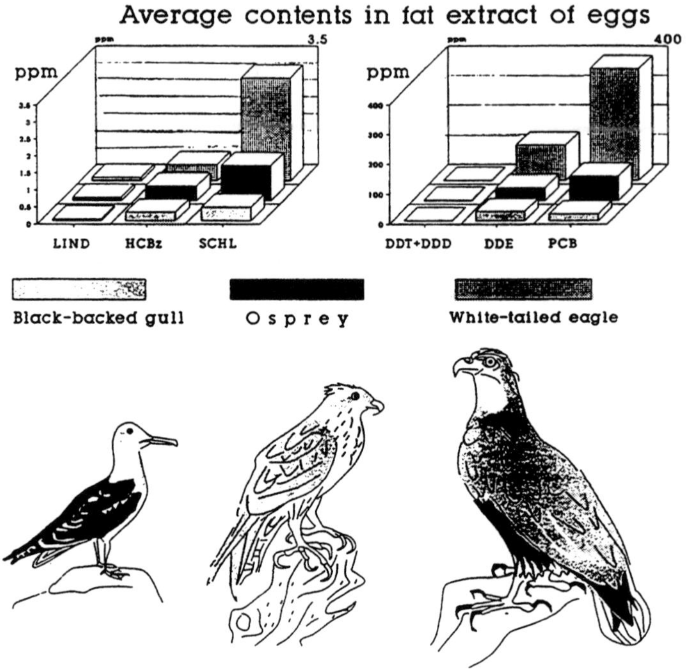


**Figure 8.3.1** Comparison of chlorohydrocarbon levels in eggs of three Baltic bird species.[10]

Nonpregnant Baltic ringed seal (*Pusa hispida*) females with pathological changes in uteri had significantly higher levels of DDT and PCB residues than normal females.[12] The same phenomenon was observed in Baltic grey seals (*Halichoerus grypus*). Females having more than 70 ppm PCB in blubber (fat) were nonfertile.[13] Extinction threatened the whole Baltic grey seal population in 1975 to 1977, but restriction of the PCB releases and supplemental feeding of seals with clean food has saved the seals. High PCB levels, however, correlate with dioxin-type toxicants PCDD and PCDF in seals, which might contribute to or be solely responsible for the pathological effects. On the other hand, similarly toxic coplanar PCB congeners (see Section 4.6) also have significant levels in the Baltic seals.[14]

## 8.4 AROMATIC HALOGENATED ETHERS

Chlorinated phenols (see Section 6.3) are waste products of chlorination processes, but they are also manufactured in large

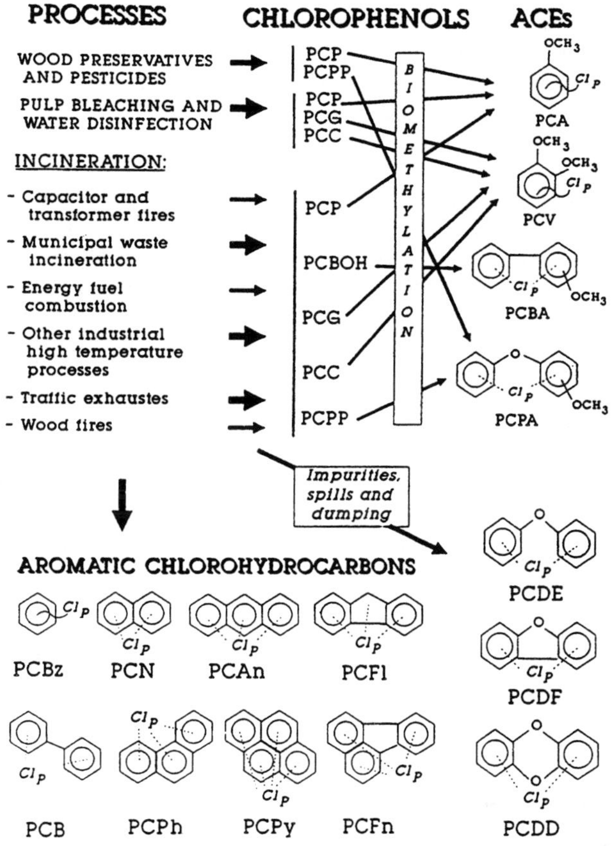


**Figure 8.4.1** Sources of environmentally important chlorophenol compounds, aromatic chloroethers (ACEs), and aromatic chlorohydrocarbons.

amounts as wood preservatives, pesticides, and hygienic agents. They are also formed in the combustion of chlorine-containing materials. Aromatic chlorohydrocarbons and ethers (ACE compounds) are formed as by-products of all these processes or by biotransformations in environment. Among the ACEs are some of the most significant hazardous substances to the environment and human health. In particular, dioxins, PCDDs and PCDFs, have become notorious to the public since 1976. A summarizing scheme of the chlorophenol sources, transformations to ACEs, and aromatic chlorohydrocarbons is shown in Figure 8.4.1 and Table 8.4.1.[2]

Dimeric chlorophenols like PCPP and PCBOH are major impurities of chlorophenol formulations. They also are inter-

**Table 8.4.1 Explanation to Figure 8.4.1**

| Abbreviation | Name | Abbreviation | Name |
|---|---|---|---|
| PCP | Polychlorophenols | PCA | Polychloroanisoles |
| PCPP | Polychlorophenoxyphenols | PCV | Polychloroveratroles |
| PCG | Polychloroguaiacols | PCBA | Polychlorobiphenylanisoles |
| PCC | Polychlorocatechols | PCPA | Polychlorophenoxyanisoles |
| PCBOH | Polychlorobiphenylols | PCDE | Polychlorodiphenyl ethers |
| PCDF | Polychlorodibenzofurans | PCDD | Polychlorodibenzo-p-dioxins |
| PCBz | Polychlorobenzenes | PCN | Polychloronaphthalenes |
| PCAn | Polychloroanthracenes | PCFl | Polychlorofluorenes |
| PCB | Polychlorobiphenyls | PCPh | Polychlorophenanthrenes |
| PCPy | Polychloropyrenes | PCFn | Polychlorofluoranthrenes |

mediates to PCDDs and PCDFs.[15,16] Incineration has been recognized as a significant source of chlorophenols and their neutral derivatives. Municipal waste incinerator ashes were reported in 1977 to contain significant levels of PCPs, PCBz's, PCBs, PCDDs, and PCDFs.[17,18]

Chlorophenols and their ethers are generally formed in all combustions of chlorine-containing materials.[19] The yields depend greatly on the temperature and the catalysts involved. In pyrolysis at 300 to 500°C and in the presence of metals like copper, aluminum, or iron (catalysts for aromatic chlorine substitution), the formation of PCDDs and PCDFs is optimal. These conditions are met in the fly ash of municipal incinerators after the combustion chamber, and a *de novo* synthesis of PCDDs and PCDFs occurs on the ash if carbon, water, chlorides, and catalytic metal traces are present.[20] Accordingly, high emissions of toxic chloroaromatics is expected from reclamations of PVC-coated electrical wires, etc.

Chlorophenols, especially PCPs and PCBOHs were found formed in air pyrolysis of PCBs at a much higher yield than PCDFs. Obviously, they are intermediates to the latter.[21] PCDEs are also found in emissions of combustions as well as impurities of chlorophenols, and they also are intermediates to PCDFs and PCDDs. PCDEs, and also the biomethylation products of chlorophenols (PCAs, PCVs, PCBAs, and PCPAs) are more persistent than their parent phenols and are frequently found in biota.[22]

Chlorophenol emissions from traffic were first indicated by

snow analyses on the roadsides, where the precursor of TCDD, 2,4,5-trichlorophenol was especially abundant.[23,24] Then, PCDDs and PCDFs were analyzed in automobile exhausts.[25] The sources were the chlorohydrocarbon additives in leaded gasoline. Accordingly, when bromohydrocarbons were used in gasoline, the exhausts contained bromine analogs and mixed bromine-chlorine analogs of chlorophenols and their ethers.

In 1987, many sources for chlorophenols and their persistent and toxic derivatives like PCDD and PCDF in the environment had been detected.[26] Bromoanalogs are formed in similar processes whenever bromine is available. Analyses of the latter are extremely complicated due to the numbers of possible congeners. However, their abundance is very much lower and their toxic significance less than that of congeners that contain only chlorine.

PCDD and PCDF in urban air particles and in sediments had very similar congener and isomer profiles. Therefore, the majority of these pollutants must have been emitted into the atmosphere from combustion sources. Because even burning of wood produces PCDDs and PCDFs, it was suggested that these pollutants have appeared in significant levels since the advent of fire. The history of PCDD/F pollution was investigated in sediment studies (Figure 8.4.2).[27,28]

Siskiwit Lake on the Isle Royale, Lake Superior, had received only atmospheric input, but sediment there contained the same combustion-related profile of PCDD/Fs as the sediments of the Great Lakes and Lake Zürich. Dating of the cores proved that accumulation of PCDD and PCDF in the environment is a recent phenomenon and related to industrial activities.[28]

## 8.5 TOXIC LOADS

### 8.5.1 Human Effects Related to Wildlife

In many cases, environmental toxicity of OHCs can be predicted from observations on accidental exposures of humans. PCDD toxicity appeared first as a severe skin disease named *chloracne* in workers of chlorine production plants. Then a series of industrial exposures from overheating in the synthesis of 2,4,5-trichlorophenol (for production of the herbicide 245-T and the hygienic compound hexachlorophene) caused chloracne

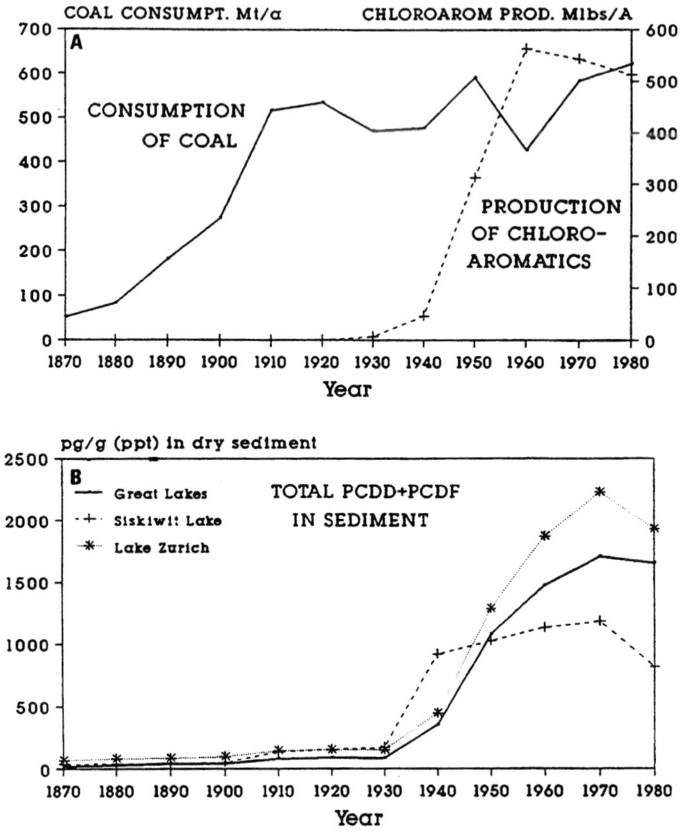


**Figure 8.4.2** (A) U.S. consumption of coal and production of synthetic chlorinated aromatic compounds. (B) Total PCDD and PCDF in sediments of three lake areas.[27,28]

and other toxic symptoms in hundreds of workers. The first well-documented case was the runaway reaction blow (explosion) in the Monsanto Nitro plant, Virginia, U.S.A., in 1949. After two German accidents, the active agent was sought and found in 1957 to be 2,3,7,8-tetra-chlorodibenzo-p-dioxin (TCDD). The TCDD blow from a factory in Seveso, Italy in 1976 notoriously contaminated the environment downwind. Within two years of the accident, 3293 animals (4.1% of the total) had died from TCDD. In humans, mainly skin symptoms were observed. Heavy uses of 245-T, especially (in Agent Orange) in the Vietnam war, and dumpings of the chlorophenol wastes have caused high local occurrences of TCDD that have been connected to the deaths of animals and to some long-term effects in people.[2,29] Other 2,3,7,8-tetrachloro-substituted PCDDs have been found to have similar toxicity (see Section 4.6) and threaten the wildlife from spills of chlorophenolic wastes and combustion emissions (see Section 8.4).[2]

After the chick edema catastrophe in 1957, heated PCBs caused mass poisoning of people in Yusho, Japan in 1968. The victims had consumed rice oil that was contaminated by leakage of the heat transfer liquid in factory. About 1600 people got severe symptoms similar to those caused by TCDD. Another heated PCB-contaminated rice oil catastrophe victimizing 1800 people occurred in Yu Cheng, Taiwan, in 1979. The main toxicants in the heated PCB were identified with PCDFs in 1970.[2]

After discovery of the dioxin receptor in 1975, not only laterally substituted PCDDs and PCDFs but also other chloroaromatic and bromoaromatic compounds having similar structures were understood to have the same toxic effects. Coplanar PCBs are also such highly toxic xenobiotics. They showed much higher enrichment rates than PCDD/Fs, especially to marine mammals and to birds.[14,30,31,32]

### 8.5.2 Load Estimation as Toxic Equivalent

To find out the most environmentally hazardous OHCs, the level of an individual persistent compound in a wildlife species (Cb) could be multiplied by the toxicity equivalent factor (TEF), which gives the level as toxic equivalents (TEQ; see Section 4.6): $TEQ = TEF \times Cb$.

Toxicology and ecotoxicology research can evaluate the TEFs for each suspected xenobiotic by structure-activity relationships (SARs), toxicity tests, and other experiences.

### 8.5.3 Dioxin and Coplanar PCB Loads

Toxic loads of PCBs could be compared to those of PCDDs and PCDFs as TEQs (TEF of 2,3,7,8-TCDD as reference; see Section 4.6). AHH-induction-based comparison of TEQs shows that coplanar PCBs represent a higher toxic load in Japanese and Finnish wildlife than PDDs and PCDFs (Table 8.5.1 and Figure 8.5.1).[14,32]

**Table 8.5.1 AHH-induction-based toxic loads ng/g of TCDD/Fs and coplanar PCBs (PCB 77+126+169+105+118+156; see Table 4.6.2) in terrestrial (human), coastal (finless porpoise), and open ocean (killer whale) mammals in Japan[32]**

| Compounds | TEQ/Human | TEQ/Finless Porpoise | TEQ/Killer Whale |
|---|---|---|---|
| PCDDs + PCDFs | 0.0706 | 0.0282 | 0.0095 |
| Coplanar PCBs | 0.369 | 4.157 | 12.84 |

Latin names: Finless porpoise = *Neophocoena phocoenoides*
Killer whale = *Orcinus orca*

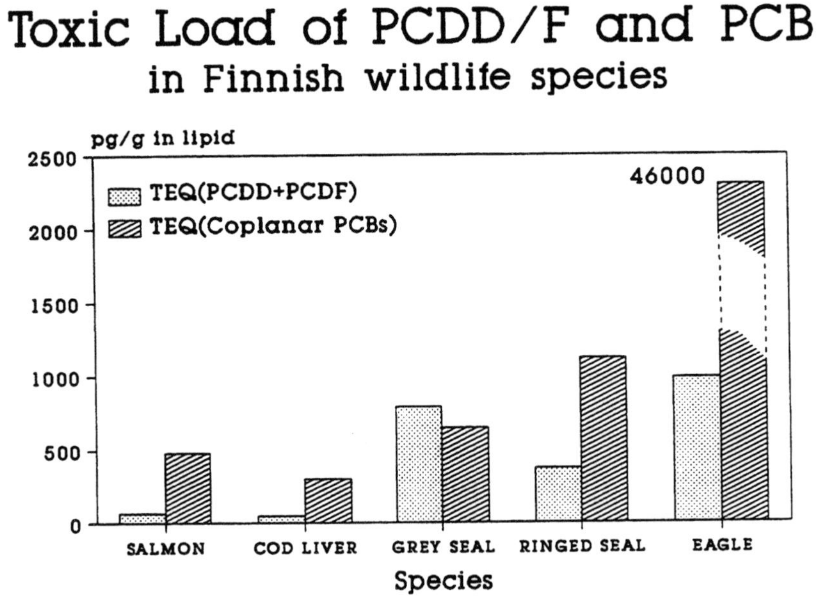


**Figure 8.5.1** AHH-based sums of TEQs for PCDD/Fs and PCBs in five different Finnish species. Coplanar PCBs included were PCB 77, PCB 105, PCB 126, and PCB 169 (see Section 4.6).

## 8.6 LOCAL AND GLOBAL OHC POLLUTION

Comparison of pollution levels in different areas can be used in estimating the nature of persistent pollutants as local, point-source related or as global, atmospherically transported.[3]

Sediment and fish studies at the Great Lakes showed that organochlorine hydrocarbons, even the rather nonvolatile PCBs, are transported long distances. Siskiwit Lake trout (see Section 8.4), which could collect only atmospherically transported xenobiotics, were good indicators of airborne organochlorines. The main pollutants in these trout were 1974 to 1976 PCBs, DDT residues, and chlordanes. In 1980, toxaphene had increased to higher levels than PCBs and DDT residues. Restrictions on PCB uses and replacement of DDT as a cotton pesticide in the southern U.S.A. by a massive use of toxaphene were the obvious reasons for these trends. Chlordanes were also rather abundant but showed no significant trend.[33]

To compare the appearance of organochlorines in Baltic and Arctic environments, fish samples from both areas were analyzed.[34] The Baltic salmon populations were from the Gulf of Bothnia, the Central Baltic Sea (Baltic Proper), and the Gulf of Finland. Arctic salmon studied were from the Tana River — their feeding area is the Arctic Ocean along the northern coast of Norway. Cod samples were from the Gulf of Finland (Baltic) and from the Vestertana Fjord at the north coast of Norway. Some results are illustrated in Figure 8.6.1.

## 8.7. SUMMARY

Persistent and toxic organohalogen compounds (OHCs) in the environment are predominantly anthropogenic. Their main sources are production for various technical purposes, pesticide uses, and industrial wastes and combustions. The toxic effects of OHCs have occurred in animals and humans in industrial accidents that have caused environmental exposure or food contamination. Also, biomagnification of OHCs has caused observable damages in wildlife, including the extinction of species.

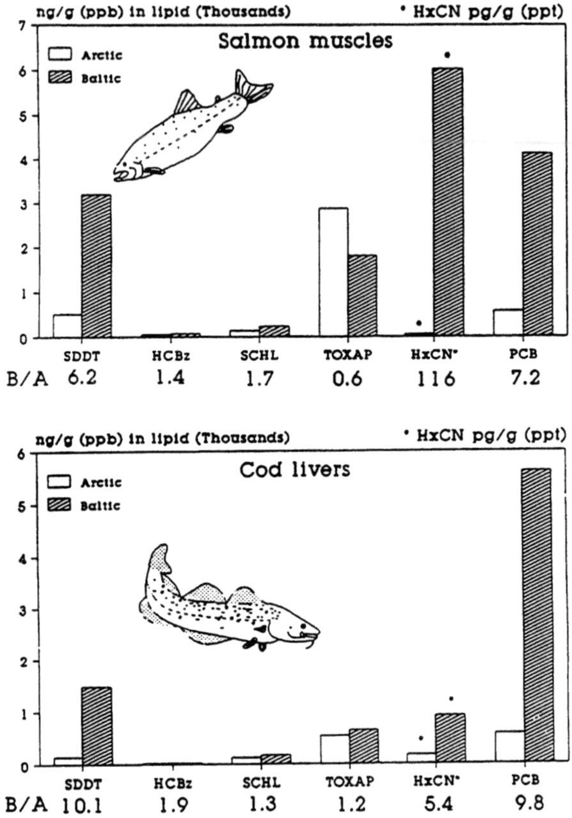


**Figure 8.6.1** Average organochlorine levels in fish samples. B/A is the ratio of concentrations in Baltic and Arctic specimens. The results indicated that hexachloronaphthalene (HxCN) is the most local and toxaphene (TOXAP) the most global of the studied pollutants.[34]

## 8.8 REFERENCES


1. **Faulkner, D. J.** "Natural Organohalogen Compounds," in *The Handbook of Environmental Chemistry, Vol. 1, Part A, The Natural Environment and Biogeochemical Cycles,* O. Hutzinger, Ed. (Berlin:Springer, 1980) pp.229-251.
2. **Paasivirta, J.** "Organochlorine Compounds in the Environment," *Water Sci.Technol.* 20:119-129 (1988).
3. **Semb, A., and J. M. Pacyna.** "Toxic Trace Elements and Chlorinated Hydrocarbons: Sources, Atmospheric Transport and Deposition," Nordic Council of Ministers, NORD 1988:74.

4. **Courtney, K. D.** "Hexachlorobenzene (HCB): A Review," *Environ. Res.* 20:225-266 (1979).
5. **Ahling, B., and A. Björseth.** "Formation of Chlorinated Substances During Combustion of PVC," *Chemosphere* 7:799-806 (1978).
6. **Matsumura, F.** *Toxicology of Insecticides* (New York: Plenum Press, 1985).
7. **Jensen, S.** "Report of a New Environmental Hazard," *New Scientist* 32:612 (1966).
8. **D'ltri, F. M., and M. A. Kamrin.** *PCBs: Human and Environmental Hazards.* (Boston:Butterworth, 1983).
9. **Lai, D. Y.** "Halogenated Benzenes, Naphthalenes, Biphenyls and Terphenyls in the Environment: Their Carcinogenic, Mutagenic and Teratogenic Potential and Toxic Effects," *J.Environ.Sci. Health* C2(2):135-184 (1984).
10. **Paasivirta, J., R. Paukku, K. Surma-Aho, J. Tarhanen, and L. Welling.** "Analyses of Environmental Toxicants in Birds" (in Finnish), Report to the Ministry of Environment, Finland 24.1.(1983).
11. **Gilbertson, M.** "Etiology of Chick Edema Disease in Herring Gulls in the Lower Great Lakes," *Chemosphere* 12:357-370 (1983).
12. **Helle, E., M. Olsson, and S. Jensen.** "PCB Levels Correlated With Pathological Changes in Seal Uteri," *Ambio* 5:261-263 (1976).
13. **Kihlström, J. E., M. Olsson, and S. Jensen.** "Effects of Organic Xenobiotics on Higher Animals" (in Swedish), in Organiska Miljögifter i Vatten. NORDFORSK Miljövårdsserien Publikation 1976:2L:567-576 (1966).
14. **Paasivirta, J., J. Koistinen, E. Mannila, J. Tarhanen, P. J. Vuorinen, N. Kannan, N. Yamashita, S. Tanabe, and R. Tatsukawa.** "Coplanar PCBs in Finnish Wildlife." Chapter 7 in *Coplanar PCBs*. R.Tatsukawa and J. P. Giesy, Eds. (Chelsea, MI:Lewis Publishers, in press).
15. **Jensen, S., and L. Renberg.** "Contaminants in Pentachlorophenol: Chlorinated Dioxins and Predioxins," *Ambio* 1(2):1-4 (1972).
16. **Paasivirta, J., M. Lahtiperä, and T. Leskijärvi.** Experiences of Structure Analyses of Chlorophenol Dimers and Trimers Found in Different Samples," in *Chlorinated Dioxins and Related Compounds*, O. Hutzinger et al. Eds. (New York:Pergamon Press, 1982) pp. 191-200.

17. **Lahaniatis,E. S., H. Parlar, and F. Korte.** "Über das vorkommen chlorierter Kohlenwasserstoffe in Flugashen von Mühlverbrennungsanlagen," *Chemosphere* 6:11-16 (1977).
18. **Olie, K., P. L. Vermeulen, and O.,Hutzinger.** "Chlorodibenzo-p-dioxins and Chlorodibenzofurans are Trace Compounds of Fly Ash and Flue Gas of Some Municipal Incinerators in The Netherlands," *Chemosphere* 6:455-459 (1977).
19. **Ahling, B. and A. Lindskog.** "Emission of Chlorinated Organic Substances from Combustion," in *Chlorinated Dioxins and Related Compounds*, O. Hutzinger et al., Eds. (New York:Pergamon Press, 1982) pp. 225-266.
20. **Stieglitz, L., and H. Vogg.** "On Formation Conditions of PCDD/PCDF in Fly Ash from Municipal Waste Incinerators," *Chemosphere* 16:1917-1922 (1987).
21. **Paasivirta, J., R. Herzschuh, T. Humppi, E. Kantolahti, J. Knuutinen, M. Lahtiperä, R. Laitinen, J. Salovaara, J. Tarhanen, and L. Virkki.** "Pyrolysis Products of PCB," *Env.Health Persp.* 60:269-278 (1985).
22. **Paasivirta, J., J. Tarhanen, and J. Soikkeli.** "Occurrence and Fate of Polychlorinated Aromatic Ethers (PCDE, PCA, PCV, PCPA and PVBA) in Environment," *Chemosphere* 15:1429-1433 (1986).
23. **Paasivirta, J., K. Heinola, T. Humppi, A. Karjalainen, J. Knuutinen, K. Mäntykoski, R. Paukku, T. Piilola, K. Surma-Aho, J. Tarhanen, L. Welling, H. Vihonen, and J. Särkkä.** "Polychlorinated Phenols, Guaiacols and Catechols in Environment," *Chemosphere* 14:469-491 (1985).
24. **Paasivirta, J., M. Knuutila, R. Paukku, and S. Herve.** "Study of Organochlorine Pollutants in Snow at North Pole and Comparison to the Snow at North, Central and South Finland," *Chemosphere* 14:1741-1748 (1985).
25. **Ballschmiter, K., H. Buchert, R. Niemczyk, A. Munder, and M. Swerev.** "Automobile Exhausts Versus Municipal Waste Incineration as Sources of the Polychloro-dibenzodioxins (PCDD) and -furans (PCDF) Found in Environment," *Chemosphere* 15:901-915 (1986).
26. **Rappe, C., R. Andersson, P-A. Bergqvist, C. Brohede, M. Hansson, L-O. Kjeller, G. Lindström, S. Marklund, M. Nygren, S. E. Swanson, M. Tysklind, and K. Wiberg.**

"Overview on Environmental Fate of Chlorinated Dioxins and Dibenzofurans. Sources, Levels and Isomeric Pattern in Various Matrixes," *Chemosphere* 16:603-1618 (1987).

27. **Czuczwa, J. M., R. and A. Hites.** "Airborne Dioxins and Dibenzofurans: Sources and Fates," *Environ. Sci. Technol.* 20:95-200 (1986).
28. **Czuczwa, M., and R. A.Hites.** "Sources and Fate of PCDD and PCDF," *Chemosphere* 15:1417-1420 (1986).
29. **Hutzinger, O., R. W. Frei, E. Merian, and F. Pocciari, Eds.** *Chlorinated Dioxins and Related Compounds. Impact on the Environment* (Oxord:Pergamon Press, 1982).
30. **Tanabe, S., N. Kannan, A. Subramanian, S. Watanabe, and R. Tatsukawa.** "Highly Toxic Coplanar PCBs: Occurence, Source, Persistency and Toxic Implications to Wildlife and Humans," *Environ. Pollut.* 47:147-163 (1987).
31. **Tarhanen, J., J. Koistinen, J. Paasivirta, P. J. Vuorinen, J. Koivusaari. I. Nuuja, N. Kannan, and R. Tatsukawa.** "Toxic Significance of Planar Aromatic Compounds in Baltic Ecosystem — New Studies on Extremely Toxic Coplanar PCBs," *Chemosphere* 18:1967-1077 (1989).
32. **Tanabe, S., N. Kannan, M. Fukushima, T. Okamoto, T. Wakimoto, and R. Tatsukawa.** "Persistent Organochlorines in Japanese Coastal Waters: An Introspective Aummary from a Far East Developed Nation," *Mar. Pollut. Bull.*, 20 :344-352 (1989).
33. **Murphy, T. J.** "Atmospheric Inputs of Chlorinated Hydrocarbons to the Great Lakes," in *Toxic Contaminants in the Great Lakes*, J. O. Ngiaru and M. S. Simmons, Eds. (New York:Wiley, 1984) pp. 53-79.
34. **Vuorinen, P. J., J. Paasivirta, and J. Koistinen.** "Comparison of the Levels of Organochlorines in Arctic and Baltic Wildlife Samples. Global Significance of the Transport and Accumulation of Polychlorinated Hydrocarbons in the Arctic, The 8th International Conference of CAI, Oslo, Norway, 18-22 September (1989).

# 9

# Emissions from the Forest Industry

## 9.1 INTRODUCTION

Pulp production in the forest industry has been one of the most serious pollution sources to the aquatic environment. However, discharges from the manufacture of unbleached pulp have been diminished very effectively by the development of recycling technology. The bleaching processes in which residual lignin is degraded are still of concern. Recycling of the bleaching chemicals is not possible and the amounts of bleaching-plant wastewater are high. Many wastes are very toxic, and some organohalogen compounds (OHCs) from bleaching with chlorine chemicals are persistent and bioaccumulating environmental toxicants. Chlorodisinfection of water produces the same harmful OHCs as pulp chlorobleaching, due to the similarity of residual lignin and extracts in nonbleached pulp and dissolved organic matter (DOM), mainly humus, in raw water and in purified tap or wastewaters to be disinfected.

Chlorobleaching of pulp with sweetwater produces organochlorines, but mills on seashores also produce some bromine compounds. This complication is due to the very easy aromatic substitution by bromine, which is formed by the oxidation of marine bromide ions with chlorine or other oxidizing bleach agents.

Water quality and pulp-mill effluents are monitored by measuring organically bound halogen: AOX (meaning adsorbable organic halogen). Due to microcoulometric titration as the final determination stage, AOX is very sensitive but does not distinguish between Cl and Br compounds.

## 9.2 LOAD PARAMETERS

The pollution load from a forest industry plant to its receiving waters is measured, e.g., by the following parameters:

- Suspended matter (SM)
- Biological oxygen demand (BOD)
- Chemical oxygen demand (COD)
- Phosphorus (P)
- Nitrogen (N)
- Total organic chlorine (TOCl)
- Adsorbable organic halogen (AOX)

In the OECD countries with major forest industries, especially the U.S.A., Canada, Sweden, and Finland, loads of SM, BOD, and COD were effectively minimized by process improvements and building up biological purification plants for the wastewaters in 1960 to 1990. Nutrients (P and N) were still problematic due to their eutrophication effect. A general goal for the 1990s was to reduce P to lower than 100 and N to lower than 300 grams per ton of pulp produced.[1]

AOX was lowered, but in 1990, it was still at a level that indicated environmental harm — chlorine chemicals were still necessary for economical production of white pulp from most softwood raw materials, and the purification plants were not able to remove all AOX.

## 9.3 ORGANOCHLORINE COMPOUNDS DISCHARGED

### 9.3.1 Amounts

Typical discharges of organochlorines from an inland kraft mill producing bleached pine pulp in 1983 are presented in Table 9.3.1.

### 9.3.2 Chlorolignin

The majority of the organochlorines, the high-molecular weight chlorolignin is less toxic than the low-molecular weight portion of the effluent. In the case of a slowly flowing recipient watercourse, most chlorolignin is deposited on the bottom within 100 km of the source (Figure 9.3.1).

In recipient rivers with faster flows, chlorolignin is transferred over longer distances, often to the sea. Environmental toxicity of chlorolignin is not significant, but its degradation to small-molecular-weight compounds might cause problems.

**Table 9.3.1 Organochlorines as dry matter (DM) and as organically bound chlorine (OCl) per ton of bleached pine pulp discharged from an inland kraft mill in 1983[2]**

| No. Compound Group | DM (g) | OCl (g) |
|---|---|---|
| 1. Chlorolignin (OCl-compounds having M > 1000) | 30,000 | 2700 |
| 2. Carboxylic acids | 200 | 60 |
| 3. Chlorophenols (PCP, PCC, PCG etc.) | 100 | 60 |
| 4. Neutral compounds: | 150 | 80 |
| 4a. Aromatic chlorohydrocarbons | 5 | 2 |
| 4b. Chloroform | 20 | 18 |
| 4c. PCDD + PCDF (as TEQs) | 2*10-6 | $10^{-6}$ |
| 4d. Other neutral compounds | 125 | 60 |
| Sum | | TOCl = 2900 |

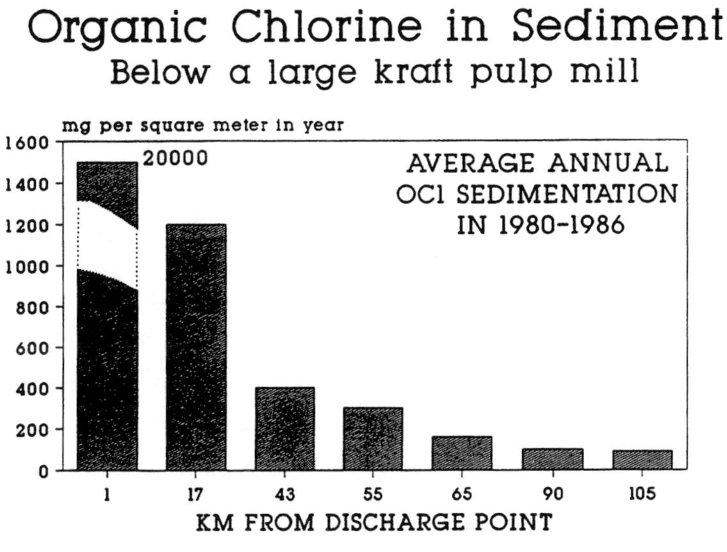


**Figure 9.3.1** OCl (chlorine bound mainly to chlorolignin) as average fallout in sediments downstream from a large kraft pulp mill in Central Finland.[3]

## 9.3.3 Carboxylic Acids

Acids in pulp-mill effluent, especially resin acids and their chlorination products, are very toxic to biota and are found enriched as conjugates in fish bile.[4,5] However, they have rather low persistence and bioaccumulation potential.[2] Therefore, their environmental hazard is limited to the immediate area near the mill.

### 9.3.4 Chlorophenols

Chlorophenols are significant components in chlorobleaching effluents. The best known of them are polychlorophenols (PCPs), guaiacols (PCGs), and catechols (PCCs).[6-11] They are very toxic and accumulate from water to biota.[12-15] Similarly to resin acids, they appear enriched as conjugates in fish bile.[14,16] Portions of them are found in water a long time after and long distances from the discharge.[15,17-22] Most of the PCGs and PCCs in sediments are chemically bound.[26] In 1990, it was observed that the majority of the PCGs and PCCs were bound in the high-molecular-weight chlorolignin already in the pulp bleachery.[56]

Other significant sources of chlorophenols in the environment are wood preservation and combustion processes (see Section 8.4). However, these produce primarily three PCP components.[15,17-19] These three compounds can be handled together as group S1PCP and the others, mainly chlorophenolics from pulp bleaching and water chlorodisinfection, as S2PCP (Figure 9.3.2).

PCGs bioaccumulate more than PCPs or PCCs (see Section 6.3), but the food-chain enrichment potential of chlorophenols is significantly lower than that of the persistent chlorohydrocarbons.[15,17-19,22]

Fish near the pulp mills seem to excrete chlorophenols faster and, therefore, do not have significantly higher content than fish from less contaminated lakes. The S2PCP components are, however, more abundant in fish polluted by the pulp mill effluents.[15,17-19,23,24]

Biomethylation of chlorophenols to more persistent and bioaccumulating polychloroanisoles (PCA) and -veratroles (PCV; see Section 6.4) is a significant cause of environmental and economic damages, because PCA and PCV are extremely bad-tasting compounds.[23,25-27] For example, in sensory panel studies of water solutions, concentration of a detectable odor lowered three to ten orders of magnitude by methylation of chlorophenols (Figure 9.3.3). The threshold taste concentrations of PCAs were so low that they could not be measured.

Analyses together with a blind-taste panel study, proved that PCAs and PCVs were the main tainting substances of fish in pulp-mill recipient waters.[24] Analyses of fish and incubated mussels showed that PCA and PCV contents correlate signifi-

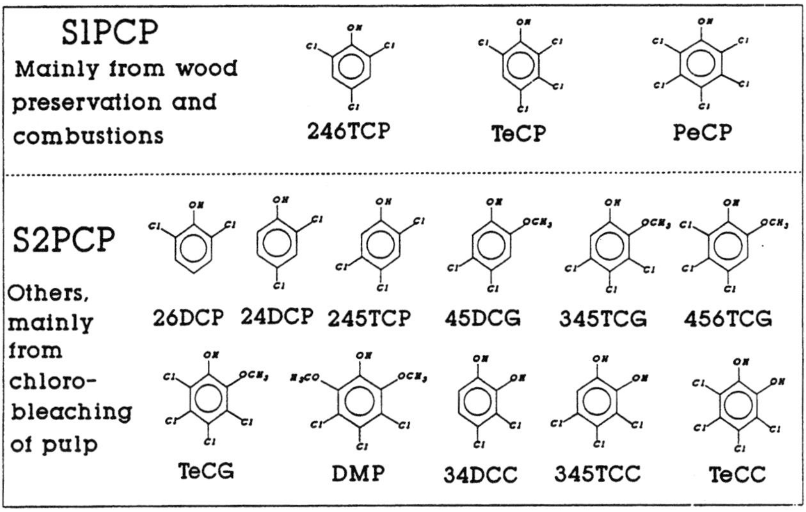


**Figure 9.3.2** The most common chlorophenols in the environment.

cantly with those of the corresponding chlorophenol compounds.[22,24] PCCs and PCGs transfer to PCVs in recipient sediments, too.[26]

## 9.3.5 Neutral Compounds

Only a minor potion of the small-molecular organochlorines in pulp bleaching effluents were characterized in 1987. Most unknown substances appeared in the neutral fraction.[29] More was known about different structure types, which could be approximately quantitated after different group separations.

One group is the extremely volatile halohydrocarbons. Its most abundant member is chloroform, which was identified in 1934.[30] Use of hypochlorite in bleaching causes chloroform to be the main fraction of neutral small-molecular organohalogen compounds in spent bleach liquor (SBL).[31] The C1 stage of bleaching (acid chlorination) produces additional substances. The C1 stage was measured to yield the following amounts of substances per ton of pine pulp bleached:[2]

| | | | |
|---|---|---|---|
| 1,2-Dichloroethane | 46 g/t | Carbon tetrachloride | 1 g/t |
| Dichloromethane | 35 g/t | 1,1,2-Trichloroethane | 1 g/t |
| Chloroform | 10 g/t | | |

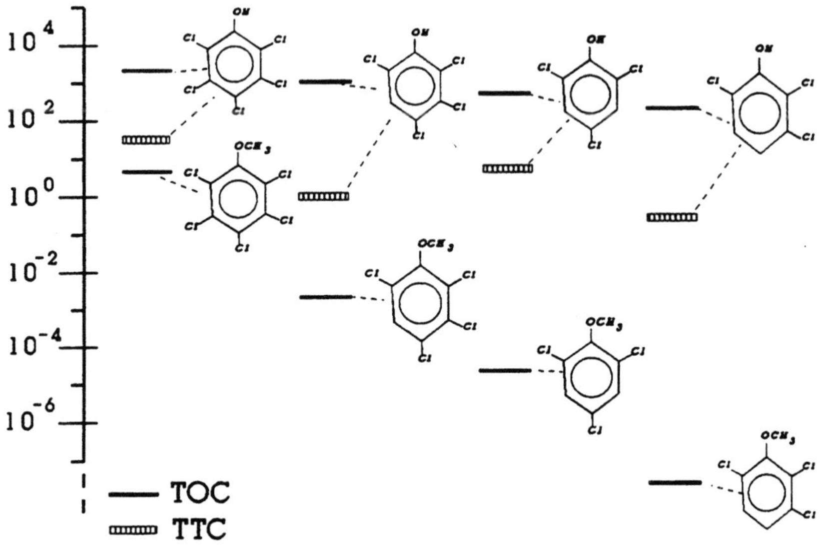


**Figure 9.3.3** Threshold odor concentrations (TOC) and threshold taste concentrations (TTC) of different chlorophenols and TOCs of the corresponding chloroanisoles.[23,27,28]

Trichloroethene and tetrachloroethene have also been found in pulp mill effluents.[31]

Chloroform was measured in significant levels (several μg/l) 100 km (two years) downstream from mill.[32] Its bioaccumulation rate in fish in natural conditions, however, was low relative to that of carbon tetrachloride, trichloroethene, and tetrachloroethene.[18] Consequently, the latter three are possibly more detrimental to food quality than chloroform as contributors to the long-term toxic load. In biological treatments such as the activated sludge process, the main portion of these volatiles disappear from wastewater, perhaps partly by evaporation.

Many of the oxygen-containing small-molecular-weight pulp-mill effluent compounds are mutagenic.[31,33,34] The strongest mutagen is the ring compound MX (Figure 9.3.4).[35] The environmental risk of these mutagens is limited to near the effluent point because they are not very persistent or bioaccumulating. More persistent are chlorosulfones, of which DDS and TDS (Figure 9.3.4) are the most abundant in waste liquor after biological treatment.[36]

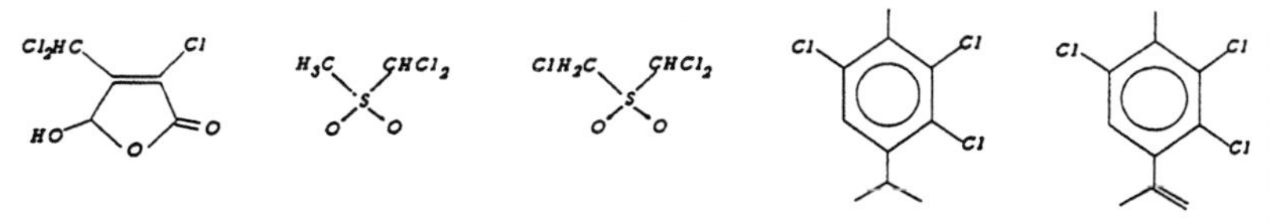

| MX | DDS | TDS | CYMS | CYMD |
|---|---|---|---|---|
| 3-Chloro-4-dichloro-methyl-5-hydroxy-2(5H)-furanone | 1,1-Dichloro-dimethyl sulfone | 1,1,1'-Trichloro-dimethyl sulfone | 2,3,6-Trichloro-p-cymene | 2,3,6-Trichloro-p-cymenene |

**Figure 9.3.4** Formula and names of five neutral compounds formed in pulp chlorobleaching.

Aromatic chlorohydrocarbons are abundant in SBL and are also found as very persistent in biologically treated wastewater. Chlorocymenes are the major components of aromatic chlorohydrocarbons in sulfite-mill effluents.[37] In contrast, chlorocymenenes are more abundant than their saturated analogues in kraft-mill waste liquors.[38,39] Two components, CYMS and CYMD (Figure 9.3.4) have been used as reference substances in environmental studies since 1980.[2] CYMS occurs at much lower levels than CYMD in kraft-mill effluents but is more abundant than the latter in biota of the recipient water-courses. CYMS is found in mussels incubated in places more than 100 km downstream from the discharge.[22]

A large number of aromatic chloroethers and thioethers are present at low levels in pulp-mill effluents.[40] PCDDs and PCDFs belong to this group of neutral compounds. Their concentrations, however, are as low as some picograms per liter in SBL and some tens of picograms per gram in pulp and paper products.[40-43] As a rule, toxic PCDDs and PCDFs are reported as TEQ values (TCDD equivalents; see Section 4.6). Annual TEQ emission into receiving waters in Sweden was estimated in 1987 to be less than 30 g.[44] This corresponds to 2 μg per ton of bleached pulp. However, dioxins were found in sediments near the pulp mill and as residues in the recipient biota.[45] Their contents in salmon and eagles showed enrichment in the food chain similar to DDT residues.[46]

A large public concern about dioxin pollution from pulp mills arose after the first report about the Environmental Protection Agency (EPA) study[43] in the U.S.A. in 1987. Emissions of toxic PCDDs and PCDFs from the forest industry, however, are

**Table 9.3.2 Estimation of annual emissions of toxic TCDDs and TCDFs as TEQs in Sweden in 1987[44]**

| Source | | TEQ g/year |
|---|---|---|
| Municipal waste incineration | (smoke gases) | 200–300 |
| Hazardous waste incineration | (smoke gases) | 2–6 |
| Car traffic | (smoke gases) | 10–150 |
| Iron and steel industry, metal factories | (smoke gases) | 100–200 |
| Coal combusting power plants | (smoke gases) | 1 |
| Hospital ovens | (smoke gases) | 10 |
| Forest industry | (smoke gases) | 4–6 |
| Forest industry | (waste water) | 30[a] |

[a] A more recent estimation given in DIOXIN'88, Umeå, was 6–7 g/year.

obviously much smaller than from many other sources (Table 9.3.2).

An estimate for the toxic equivalent (TEQ) per ton of TOCl emitted is 0.44 mg/t.[2] The annual load to the Gulf of Bothnia from all pulp mills in Sweden and Finland was estimated in 1985 to be 14,400 tons of TOCl.[47] Accordingly, the TEQ load was 6.3 g in the year 1985 and less than 300 g during the whole history of the chemical forest industry. On the other hand, TEQ in Baltic sediments was estimated to be 30,000 g.[44] This means that 1% of the toxic TCDD/Fs as TEQ in the Baltic Sea originated from the Swedish and Finnish forest industry.

The main neutral planar aromatic chlorocompounds (PLACs) from chlorobleaching are substances other than PCDDs or PCDFs. A series of compounds that interfered in GC/MS determinations of TCDD were found in sediments,[48,49] sludge,[49] and active sludge treatment plant effluent[46] of pulp mills and also in paper products.[46,50] These compounds were named UPACs (unknown planar aromatic chlorocompounds).[46] Structures for the UPACs have been suggested to be chlorinated xanthenes of xanthones,[48] alkylpolychlorodibenzofurans (R-PCDFs),[49,50] and alkylpolychlorobibenzyls (R-PCBBs).[57,58]

The UPACs follow PCDDs and PCDFs in analytical cleanup. While their emissions are about three orders of magnitude higher than those of toxic PCDD/Fs, they might contribute to dioxin-type toxic loads and environmental effects from pulp mills.

**Figure 9.4.1** Typical fin injuries in perch and an upper jaw deformation ("pug-head") in pike caught near the discharge of bleaching kraft mill in Norrsundet on the Gulf of Bothnia coast in Sweden, 1984.[51]

## 9.4 BIOLOGICAL EFFECTS OF CHLOROBLEACHING EFFLUENTS

In addition to the obvious acute toxicity of bleach kraft mill effluent (BKME) components, developmental damages, growth disturbances, and changes in population densities have been observed at further distances from the mills than where acute toxic levels were measured.[51] Examples of malformations in fish near the discharge point of the Norrsundet mill in Sweden in 1984 are shown in Figure 9.4.1.

In 1984, the Norrsundet mill employed chlorobleaching without biological purification. The 2,3,7,8-TCDD levels in perch near the discharge were 2.6-19 pg/g in fresh fish.[45] These are unusually high compared to the levels of TCDD in fish near other pulp mills. In contrast, 2,3,7,8-TCDF concentrations 2.1-8.8 pg/g were more normal Baltic background concentrations, although TCDF, as a rule, is the main PCDD/PCDF congener formed in chlorobleaching.[42,43,48]

TCDD and TCDD+BKME exposures caused a very significant, while BKME caused a nonsignificant, induction of cyto-

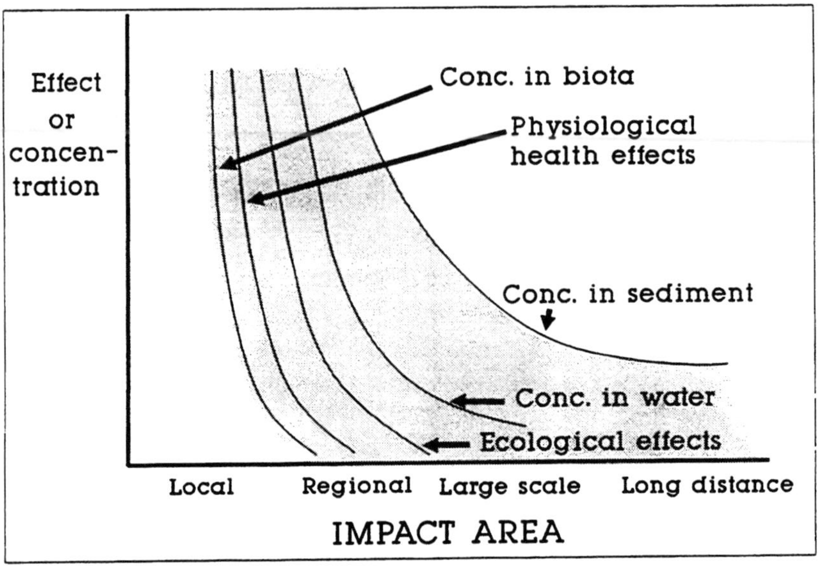


**Figure 9.4.2** Principle relationships between different observations of influence from bleached kraft pulp mills on the Gulf of Bothnia. Concentrations of organic chlorine compounds are compared with effects.[51]

chrome P-450· enzymes (AHH-related pentoxyresorufin-O-dealkylase [PROD] and ethoxyresorufin-O-dethylase [EROD]) in fish livers.[52] This result indicates that the role of the TCDD and its analogues in toxicity of bleached kraft mill discharge is normally insignificant.

Mussel incubations in Finland during the summer of 1988 resulted in marked accumulation of toxic PCDD+PCDF congeners only in the Kymi River where kraft mills did not have any biological purification plant.[53] The following summer (1989) the TEQ load of PCDD/Fs in mussels incubated in the Kymi River decreased tenfold, obviously due to the start of a new activated sludge treatment plant there.

A summarizing picture of the extent of measurable elevated concentrations of organochlorine compounds related to extent of the observed effects is given in Figure 9.4.2.

In freshwater fish exposed to kraft pulp mills, physiological disturbances are also evident. However, the effects were small and a considerable development of increased tolerance was observed.[54] Organochlorines deposited on bottoms at long distances from the mills might contribute to the malformations in benthic fauna that were observed in exposure experiments.[55]

## 9.5 RISK ASSESSMENT

### 9.5.1 Calculation of the Relative Risk

To minimize the toxic load of pulping upon the environment, it is useful to know relative risk levels of different organochlorine substances in BKME. For comparisons, the relative risk $R_i$ for a specified consequence or damage caused by the compound group *i* is calculated (as presented in Section 3.1):

$$R_i = \mathrm{Ex} \times \mathrm{Ef},$$

where Ex = exposure and Ef = effect in the risk situation.

### 9.5.2 Deaths of Exposed Fish in Recipient

To calculate $R_i$'s in the case of lethal effect of BKME to fish at the discharge point = $R_f$'s, Ex values are taken from relative emissions and Ef's as inverses of average $LC_{50}$ values (toxicities) of the organochlorine compound groups compared:

$$R_f = \mathrm{Ex} \times 1 / LC_{50}$$

Values are presented in Table 9.5.1.

As a conclusion, chlorophenols represent the greatest lethal risk to fish near the mill.

### 9.5.3 Risk that Food is Inedible

Pollutants can bioaccumulate in finfish, grabs and shellfish to concentrations that make them inedible. The effect might be tainting (bad taste) or a health ban on the sale or consumption of seafood. To evaluate the inedibility risk ($R_n$), emission is multiplied by the average bioconcentration factor of the compound group (Bcf). The effect factor is expressed as the inverse threshold concentration or acceptable level (1/AL):

$$R_n = \mathrm{Ex} \times \mathrm{Bcf} / \mathrm{AL}.$$

Values are presented in Table 9.5.2.

**Table 9.5.1 Relative risk $R_f$ to fish death in recipient water near the discharge point of BKM**

| Compound Group | Ex = Emission mg/t | $LC_{50}$ mg/l | $R_f$ l/t |
|---|---|---|---|
| Chlorophenols | 100,000 | 0.5 | 200,000 |
| Aromatic chlorohydrocarbons | 5,000 | 1.7 | 3,000 |
| Chloroform | 20,000 | 2.0 | 10,000 |
| PCDD + PCDF as TEQ | 0.002 | .000005 | 400 |

**Table 9.5.2 Evaluation of the relative risk factors ($R_n$) for inedibility of the recipient fish**

| Compound Group | Bcf | Basis of the Acceptable Level | AL μg/kg | $R_n$ |
|---|---|---|---|---|
| Chlorophenols | 475 | Bad taste[24] | 100 | 475,000 |
| Arom.CHCs (HCBz model) | 36,000 | Food regulation[a] | 2,000 | 90,000 |
| Chloroform | 9 | Drinkwater reg[a] | 1.9 | 95,000 |
| PCDD + PCDF as TEQ | 240,000 | FDA regul.1981 | 0.050 | 9,600 |

[a] AL values set by Finnish authorities.

Chlorophenols remained as the greatest risk in this case, too. Actually, the tainting compounds in fish were PCAs and PCVs, which correlated with the content of their precursor chlorophenols in fish.[24]

## 9.5.4 Long-Term Environmental Risk

For persistent effluent components, a long-term exposure of animals and humans at the top of the food chains can be calculated from both bioconcentration (Bcf) and biomagnification factors (Bmf). In the present estimation, Bmf = ratio of the average concentrations in white-tailed eagles ($C_e$) and salmon ($C_s$) in the Central Gulf of Bothnia area in 1985. As an effect value, the inverse of the $LD_{50}$ (oral rate) is used. Chloroform is omitted from consideration as too nonpersistent in biota:

$$R_e = \mathrm{Ex} \times \mathrm{Bcf} \times \mathrm{Bmf} \,/\, \mathrm{LD}_{50}.$$

Values are in Table 9.5.3.

**Table 9.5.3 Evaluation of the long-term environmental risk factors $R_e$ for three groups of BKME components**

| Compound Group | Ex mg/t | Bcf | Bmf ($C_e/C_s$) | $LD_{50}$ mg/kg | $R_e$ |
|---|---|---|---|---|---|
| Chlorophenols | 100,000 | 475 | 1.25 | 250 | 237,500 |
| Arom. CHCs (HCBz model) | 5,000 | 36,000 | 11.8 | 5 | 4,248,000 |
| PCDD + PCDF as TEQ | 0.002 | 240,000 | 27.0 | 0.1 | 129,600 |

Consequently, aromatic chlorohydrocarbons emitted from pulp mills appear to present larger long-term environmental risk than chlorophenols or dioxins and furans from the same source. However the present calculation did not include other PLACs, which are more abundant than PCDDs and PCDFs in the effluent, because their exact nature, environmental fate, and toxicities were too ambiguous for such risk estimation in 1990.[2]

## 9.6 SUMMARY

While other emissions from the forest industry to the recipient waters have been minimized, organic chlorocompounds from chlorobleaching of pulp continue to be of ecotoxic concern. Most of the organically bound chlorine is in the high-molecular-weight chlorolignin, which is persistent and deposits on bottoms but appears not to be very hazardous to biota. Both chlorinated and nonchlorinated small-molecular-weight waste components include highly toxic substances. Of these, chlorophenol compounds appear to make the largest acute risk. However, the persistent aromatic chlorohydrocarbons and ethers represent a more long-term risk to wildlife and humans.

## 9.7 REFERENCES

1. "Seminar on the Pulp and Paper Industry," Helsinki and Paris Comissions, Stockholm, April 4-5 (1989).
2. **Paasivirta, J.** "Organochlorines in Finnish and Baltic Environment. The role of the Forest Industry," in *Biotransformation of Organic Pollutants in the Aquatic Environment*, CEC Water Pollution Research Report 14:1-10 (1989).

3. **Paasivirta, J., J. Knuutinen, P. Maatela, R. Paukku, J. Soikkeli, and J. Särkkä.** "Organic Chlorine Compounds in Lake Sediments and the Role of the Chlorobleaching Effluents," *Chemosphere* 17:137-146 (1988), and nonpublished results.
4. **Holmbom, B., and K-J. Lehtinen.** "Acute Toxicity to Fish of Kraft Pulp Mill Bleach Plant Effluents," *Paperi ja Puu* 62(11):673-684 (1980).
5. **Oikari, A., B. Holmbom, E. Ånäs, and H. Bister.** "Distribution in a Recipient Lake and Bioaccumulation in Fish of Resin Acids from Kraft Pulp Mill Waste Waters," *Paperi ja Puu* 62(4a):93-202 (1980).
6. **Dence, C. W., and K. Sarkanen.** "A Proposed Mechanism for the Acidic Chlorination of Softwood Lignin," *Tappi* 43:87-96 (1960).
7. **Servizi, J. A., R. W. Gordon, and D. W. Martens.** "Toxicity of Two Chlorinated Catechols, Possible Components of Kraft Pulp Mill Waste," Int. Pacific Salmon Fish. Comm. Progr. Report 17 (1968).
8. **Rogers, I. H.** "Isolation and Chemical Identification of Toxic Compound of Kraft Mill Wastes," *Pulp Paper Can.* 33:1858-1860 (1973).
9. **Rogers, I.H., and L. H. Keith.** "Organochlorine Compounds in Kraft Bleaching Wastes. Identification of Two Chlorinated Guaiacols," Technical Report 405, Fisheries and Marine Service, Research and Development Directorate, Pacific Environment Institute, West Vancouver, B.C. (1974).
10. **Lindström, K., and J. Nordin.** "Gas Chromatography — Mass Spectrometry of Chlorophenols in Spent Bleach Liquors," *J. Chromatogr.* 128 :13-26 (1976).
11. **Knuutinen, J.** "Synthesis, Structure Verification and Gas Chromatographic Determination of Chlorinated Catechols and Guaiacols Occuring in Spent Bleach Liquors of Kraft Pulp Mills," Department of Chemistry, University of Jyväskylä, Research Report 18 (1984).
12. **Ahlborg, U. G., and T. M. Thunberg.** "Chlorinated Phenols: Occurence, Toxicity, Metabolism, and Environmental Impact," *CRC Crit.l Rev. Toxicol.* 1-35 (July 1980).

13. **Paasivirta, J., M. L. Hattula, J. Särkkä, J. Janatuinen, M. Pitkänen, and T. Kurkirinne.** "On the Analysis and Appearance of the Organic Chlorine Compounds in the Lake Päijänne Ecosystem," NORDFORSK Miljövårdsekreteriatet 1976:2 :439-462 (1976).
14. **Kobayashi, K.** "Metabolism of Pentachlorophenol in Fishes," in *Pentachlorophenol Chemistry, Pharmacology and Environmental Toxicology*. K. R. Rao, Ed. Environmental Science Research, Vol. 12 (New York:Plenum Press, 1978) pp. 143-150.
15. **Paasivirta, J., J. Särkkä, T. Leskijärvi, and A. Roos.** "Transportation and Enrichment of Chlorinated Phenolic Compounds in Different Aquatic Food Chains," *Chemosphere* 9:441-456 (1980).
16. **Oikari, A., and B. Holmbom.** "Assesment of Water Contamination by Chlorophenolics and Resin Acids with the Aid of Fish Bile Metabolites," *Aquat. Tox. Environ. Fate.* 9:252-267 (1986).
17. **Paasivirta, J.** "Enrichment of chlorobleaching residues in food chain," NORDFORSK Miljövårdsserien 1981:1:187-195 (1981).
18. **Paasivirta, J., J. Särkkä, K. Surma-Aho, T. Humppi, T. Kuokkanen, and M. Marttinen.** "Food Chain Enrichment of Organochlorine Compounds and Mercury in Clean and Polluted Lakes of Finland," *Chemosphere* 12:239-252 (1983).
19. **Paasivirta, J., K. Heinola, T. Humppi, A. Karjalainen, J. Knuutinen, K. Mäntykoski, R. Paukku, T. Piilola, K. Surma-Aho, J. Tarhanen, L. Welling, and H. Vihonen.** "Polychlorinated Phenols, Guaiacols and Catechols in Environment," *Chemosphere* 14:469-491 (1985).
20. **Paasivirta, J., J. Knuutinen, P. Maatela, R. Paukku, J. Soikkeli, and J. Särkkä.** "Organic Chlorine Compounds in Lake Sediments and the Role of the Chlorobleaching Effluents," *Chemosphere* 17:137-146 (1988).
21. **Paasivirta, J., J. Knuutinen, M. Knuutila, P. Maatela, O. Pastinen, L. Virkki, R. Paukku, and S. Herve.** Lignin and Organic Chlorine Compounds in Lake Water and the Role of the Chlorobleaching Effluents," *Chemosphere* 17:147-158 (1988).

22. **Herve, S., P. Heinonen, R. Paukku, M. Knuutila, J. Koistinen, and J. Paasivirta.** "Mussel Incubation Method for Monitoring Organochlorine Pollutants in Watercourses. Four-year application in Finland," Chemosphere 17:1945-1961 (1988).
23. **Paasivirta, J., J. Knuutinen, J. Tarhanen, T. Kuokkanen, K. Surma-Aho, R. Paukku, H. Kääriäinen, M. Lahtiperä, and A. Veijanen.** "Potential Off-Flavour Compounds from Chlorobleaching of Pulp and Chlorodisinfection of Water," *Wat. Sci. Technol.* 15:97-104 (1988).
24. **Paasivirta, J., P. Klein, M. Knuutila, J. Knuutinen, M. Lahtiperä, R. Paukku, A. Veijanen, L. Welling, M. Vuorinen, and P. J. Vuorinen.** "Chlorinated Anisoles and Veratroles in Fish. Model Compounds. Instrumental and Sensory Determinations," *Chemosphere* 16:1231-1241 (1987).
25. **Paasivirta, J., J. Tarhanen, and J. Soikkeli.** "Occurrence and Fate of Polychlorinated Aromatic Ethers (PCDE, PCA, PCV, PCPA and PCBA) in Environment," *Chemosphere* 15:1429-1433 (1986).
26. **Remberger, M., A-S. Allard, and A. H. Neilson.** "Biotransformations of Chloroguaiacols, Chlorocatechols and Chloroveratroles in Sediments," *Appl. Environ. Microbiol.* 51(3):552-558 (1986).
27. **Curtis, R. F., D. G. Land, N. M. Griffiths, M. Gee, D. Robinson, J. L. Peel, C. Dennis, and J. M. Gee.** "2,3,4,6-Tetrachloroanisole Association with Musty Taint in Chickens and Microbial Formation," *Nature* 235: 223-224 (1972).
28. **Dietz, F., and J. Traud.** "Geruch- und Geschmacks-Schwellen-Konzentrationen von Phenolkörpern," *GWF-Wasser/Abwasser* 119 H ,6:318-325 (1978).
29. **Suntio, L. R., W. Y. Shiu, and D. Mackay.** "A Review of the Nature and Properties of Chemicals Present in Pulp Mill Effluents," **Chemosphere** 17:1249-1290 (1988).
30. **Harris, E. E., E. C. Sherrard, and R. L. Mitchell.** "Some Chemical Reactions of Maple and Spruce Lignin," *J.Am. Chem. Soc.* 56:889-893 (1934).
31. **Stockman, L., L. Strömberg, and P. de Sousa.** "Mutagenic Properties of Bleach Plant Effluents: Present Stage of Knowledge," *Cellulose Chem.Technol.* 14:517-526 (1980).

32. **Toivanen, E., and H. Harjula.** "Förekomst av trihalometaner och klorerade fenoler i Päijänne," NORDFORSK Miljövårdsserien 1981:1:175-185 (1981).
33. **Voss, R. H., J. T. Wearing, R. D. Mortimer, T. Kovacs, and A. Wong.** "Chlorinated Organics in Kraft Bleachery Rffluents," *Paperi ja Puu - Papper och Trä* 62(12):809-814 (1980).
34. **Kringstad, K. P., and K. Lindström.** "Present Knowledge on the Organic Chemical Composition of Spent Bleach Liquors with Emphasis on Compounds with Toxic Effects," TAPPI Proc. (1982) 191-200 (1982).
35. **Holmbom, B., R. Voss, R. D. Mortimer, and A. Wong.** "Isolation and Identification of an Ames-Mutagenic Compound Present in Kraft Chlorination Effluents," *Tappi* 64(3):172-174 (1981).
36. **Voss, R. H.** "Chlorinated Neutral Organics in Biologically Treated Bleached Kraft Mill Effluents," *Environ. Sci. Technol.* 17(9):530-537 (1983).
37. **Björseth, A. G. Lunde, and N. Gjös.** "Analysis of Organochlorine Compounds in Effluents from Bleacheries by Neutron Activation Analysis and Gas Chromatography/Mass Spectrometry," *Acta Chem. Scand.* B 31:797-801 (1977).
38. **Kuokkanen, T., and J. Paasivirta.** "Determination of Chlorocymenes in Spent Bleach Liquor from Different Bleaching Stages of a Sulphate Plant," *Kemia-Kemi* 7:763 (1980).
39. **Kuokkanen, T., and J. Koistinen.** "Contents of Aromatic Chlorohydro-carbons, Toxaphene, PCDDs and PCDFFs in Spent Bleach Liquor after Biological Purification," *Chemosphere* 19:727-230 (1989).
40. **Tarhanen, J., J. Paasivirta, J. Knuutinen, T. Humppi, R. Laitinen, and E. Kantolahti** "Model Compound Studies and Structure Analyses of Chlorophenol Dimers and Trimers which are Formed in Chlorobleaching of Pulp," 67th Annual CIC Conference, Montreal, June 3-6, (1984). Abstracts EN 2-6.
41. **Beck, H., K. Eckart, W. Mathar, and R. Wittkowski.** "Occurence of PCDD and PCDF in Different Kinds of Paper," *Chemosphere* 17:51-57 (1988).

42. **Swanson, S. E., C. Rappe, J. Malmström, and K. P. Kringstad.** "Emissions of PCDDs and PCDFs from the Pulp Industry," *Chemosphere* 17:681-691 (1988).
43. **Amendola, G., D. Barna, R. Blosser, L. LaFleur, A. McBride, F. Thomas, T. Tiernan, and R. Whittemore.** "The Occurence and Fate of PCDDs and PCDFs in Five Bleached Kraft Pulp and Paper Mills," *Chemosphere* 18:1181-1188 (1989).
44. "DIOXIN," Working group of Naturvårdsverket, Sweden, Remissupplag May (1987).
45. **Rappe, C., P-E. Bergqvist, and L-O. Kjeller.** "Levels, Trends and Patterns of PCDDs and PCDFs in Scandinavian Environmental Samples," *Chemosphere* 18:651-658 (1989).
46. **J. Paasivirta, K. Mäntykoski, J. Koistinen, T. Kuokkanen, E. Mannila, and K. Rissanen.** "Structure Analyses of Planar Polychloroaromatic Compounds in Environment," *Chemosphere* 19:149-154 (1989).
47. "Water Pollution Problems of Pulp and Paper Industries in Finland and Sweden," Committee for the Gulf of Bothnia, Report of the special working group, May (1987); Mimeograph of the Board of Waters and Environment, Finland 2 (1987).
48. **Kuehl, D. W., B. C. Butterworth, W. DeVita, and C. P. Sauer.** "Environmental Contamination by Polychlorinated Dibenzo-p-dioxins and Dibenzofurans Associated with Pulp and Paper Mill Discharges," Biomed. Environ. Mass Spectrom. 14:443-447 (1987).
49. **Buser, H-R., L-O. Kjeller, S. E. Swanson, and C. Rappe.** "Methyl-polymethyl, and Alkylpolychlorodibenzofurans Identified in Pulp Mill Sludge and Sediments," *Environ. Sci. Technol.* 23:1130-1137 (1989).
50. **Beck, H., A. Dross, K. Eckart, W. Mathar, and R. Wittkowski.** "PCDDs, PCDFs and Related Compounds in Paper Products," *Chemosphere* 19:655-660 (1989).
51. **Södergren, A., Ed.** "Biological Effects of Bleached Pulp Mill Effluents," Final Report from the Environment/Cellulose I Project, Naturvårdsverket, Sweden (1989).
52. **Lindström-Seppä P., and A. Oikari.** "Effects of TCDD on Biotransformation Responses in Fish: Studies on Rainbow Trout (Salmo gairdneri) Exposed to TCDD and Bleached

Kraft Pulpmill Effluents," *Nordic Symposium on Organic Environmental Chemicals* 11-14.9.1989, Abstracts, A. Oikari, Ed., University of Joensuu, Faculty of Mathematics and Natural Sciences Report 29:47-49 (1989).

53. **Paasivirta, J., S. Herve, P. Heinonen, and T. Rantio.** "Bioaccumulating Organochlorine Compounds in Finnish Watercourses Surveyed by the Mussel Incubations," Nordic Symposium on Organic Environmental Chemicals 11-14.9.1989, Abstracts, A. Oikari, Ed., University of Joensuu, Faculty of Mathematics and Natural Sciences Report 29:65-66 (1989).
54. **Tana, J., K-J. Lehtinen, and C. Monfelt.** "Health Status of a Fresh-Water Fish Population Downstream of a Bleached Kraft Mill," SECOTOX, Poceedings of the 1st European Conference on Ecotoxicology, Copenhagen (1989) pp. 149-150.
55. **Ekelund, R., Å. Granmo, and M. Berggren.** "Early Development of Mussels (Mytilus edulis) Influenced by Sediment Contaminated by Bleached Pulp Wastes," Nordic Symposium on Organic Environmental Chemicals 11-14.9.1989, Abstracts, A. Oikari, Ed., University of Joensuu, Faculty of Mathematics and Natural Sciences Report, 29:10 (1989).
56. **Paasivirta, J., H. Hakala, J. Knuutinen, T. Otollinen, J. Särkkä, L. Welling, R. Paukku, and R. Lammi.** "Organic Chlorine in Lake Sediments. III. Chlorohydrocarbons, Free and Chemically Bound-Chlorophenols," *Chemosphere* 21:1355–1370 (1990).
57. **Nevalainen, T., and J. Koistinen.** "Model Compounds Synthesis for the Structure Determination of New Unknown Planar Aromatic Compounds Originating from Pulp Mill," DIOXIN'90, 10th International Meeting, Bayreuth, F.R.G. 10-14.9.1990, Short Papers 4:361–364 (1990).
58. **Koistinen, J., R. Herzschuh, and J. Stach.** "PCDDs, PCDFs and Some Unknown Chlorinated Planar Compounds in Pulp Mill Products, Effluents, Sediments and Exposed Biota," DIOXIN'90, 10th International Meeting, Bayreuth, F.R.G. 10-14.9.1990, Short Papers 3:259-262 (1990).

# 10

# Oil Residues in the Environment

## 10.1 INTRODUCTION

Anthropogenic oil residues related to the environment originate from petroleum (crude oil). Petroleum is a naturally occurring complex mixture of compounds of carbon, hydrogen, sulfur, nitrogen, oxygen, and other elements. It is formed by the degradation of biota. It is present in the earth crust as natural asphalts (solids), crude oils (liquids), and gases. The known crude oil resources in 1966 were 400 billion ($4 \times 10^{11}$) barrels (one barrel = 181.7 liters).[1] Since that time, both the known resources and the production of fuels from crude oil have increased significantly.

A major risk of environmental damage is associated with the use and transportation of crude oil and oil products. The size of the risk can be estimated considering that the world capacity of oil tankers is about 200 million tons dw.[1] In 1969, the size of the tankers had grown to 300,000 tons dw and ships of 500,000 tons dw were planned.[2]

## 10.2 COMPOSITION OF THE CRUDE OILS

Petroleums consist of many thousands of different compounds. Crude oils differ greatly depending on the area of their origin. However, their main components are hydrocarbons and their derivatives. Sulfur and nitrogen compounds are more abundant than oxygen compounds. The crudes also contain detectable amounts of nickel, vanadinium, iron, and other metals.[1-3] Elemental ratios useful in identification of crude oils are listed in Table 10.2.1.

All crude oils can be divided into three main fractions of organics: alkanes, cycloalkanes, and aromatic compounds. Olefins are practically absent in crudes, but they are abundant

**Table 10.2.1 Elemental composition of different crude oils.[4]**

| Origin of the Oil | Sulfur % | Nitrogen % | Ni mg/kg | V mg/kg |
|---|---|---|---|---|
| Zarzaitine, Algeria | 0.06 | 0.02 | 2 | 5 |
| Nigeria | 0.16 | 0.09 | 12 | 3 |
| Zelten Brega, Libya | 0.24 | | 10 | 3 |
| Es Sider, Libya | 0.42 | 0.16 | 18 | 32 |
| Bai Hassan, Iraq | 1.36 | 0.28 | | 52 |
| Gach Saran, Iran | 1.57 | 0.35 | 49 | 148 |
| Tia Juana, Venezuela | 1.61 | 0.24 | 25 | 171 |
| Kuwait | 2.59 | 0.16 | 18 | 49 |
| Bachaquero, Venezuela | 2.64 | 0.41 | 68 | 426 |
| Arabian heavy | 3.00 | 0.71 | 27 | 84 |
| Boscan, Venezuela | 5.10 | 0.78 | 92 | 12 |
| Amposta, Spain | 5.70 | 0.18 | 19 | 12 |

in many petroleum products after formation via the cracking process.[2-4] In the alkane fraction, normal paraffins are found throughout the whole boiling range of petroleum.[1] Individual n-alkanes (C–n) are readily analyzed as sharp peaks in a gas chromatograph equipped with a FI-detector. Branched hydrocarbons, as a rule, represent the major mass of the alkane fraction. Most of them are nonseparable to individuals due to their myriads of forms. Some branched alkanes, however, have been characterized and are seen as individual peaks in gas chromatography (GC). Especially characteristic components are pristane and phytane (PRIS and PHY; see Figure 10.2.1).[4-7]

PRIS and PHY are produced by plankton from isoprenoids. They are abundant in young sediments where continuous formation of new oil takes place. Both resist bacterial biodegradation more than n-alkanes. PHY is more persistent than PRIS. Consequently, PRIS:PHY ratios are used to evaluate age and origin of the environmental oil contaminants.[4-7]

The most common aromatic hydrocarbons in crude oils are alkyl-substituted benzenes, naphthalenes, and phenanthrenes. Alkylated and non-alkylated anthracene, pyrene, fluoranthene, benzo(a)anthracene, chrysene, triphenylene, benzo(g,h,i)-fluoranthene, benzo(b)fluoranthene, benzo(k)fluoranthene, benzo(a)pyrene, benzo(e)perylene, and benzo(g,h,i)perylene are other characteristic constituents. Alkylated homologues are more abundant than their parent aromatic hydrocarbons in crude oils.[3,8]

Most of the sulfur-, nitrogen- and oxygen-containing com-

Pristane (PRIS) $C_{19}H_{40}$ Phytane (PHY) $C_{20}H_{42}$

**Figure 10.2.1** Structures of pristane and phytane.

pounds in crude oil are aromatic compounds. The most important sulfur compounds are alkyl and cycloalkyl thiols, sulfides, and the hetero-aromatics: alkylated thiophenes, thienyl sulfides, benzothiophenes, dibenzothiophenes and benzonaphthothiophenes.[1-3] The best-known nitrogen compounds in crudes are pyridine, quinoline, tetrahydroquinoline, benzoquinoline and dihydropyridine derivatives, carbazoles, porphyrines, pyrroles, indoles, isoquinolineacridines, carbazoquinolines, hydroxybenzoquinolines, 1,10-phenanthroline, and tetrahydrocarbazolenine. In refining processes, anilines, phenazines and nitriles are formed in oil products.[1] Oxygen compounds in petroleum include carboxylic acids, phenols, ketones, dibenzofurans, lactones, esters, and alcohols.[1]

## 10.3 FATE OF OIL IN THE MARINE ENVIRONMENT

Most crude oils and oil products have specific gravities less than one. Therefore, in case of a spill at sea, they will not readily sink after initial release but spread first as a slick on the surface of the sea. The weathering processes (Figure 10.3.1) immediately attack the slick and influence the properties of the residual oil and, consequently, its fate.[7]

On the average, 50% of the original mass of a oil slick evaporates. The degree of evaporation depends on the ambient temperatures.

Immediately after the spill, oil-in-water (O/W) emulsification takes place and the rates of degradation by photo-, microbial-, and auto-oxidation are maximal. Oil components are dissolved in water to some ppm levels, and fine O/W emulsions can contain some additional ppms.[7]

The mechanical stress of wind and waves promotes formation of an water-in-oil (W/O) emulsion called mousse. The water content of mousse is 10 to 80% (w/w) depending on the

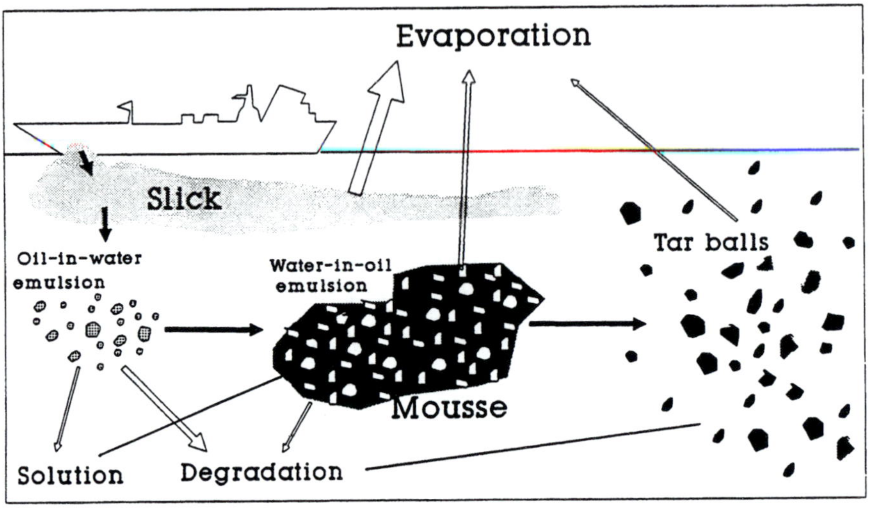


**Figure 10.3.1** The principal weathering processes of oil in the sea.

oil properties and environmental conditions. In ice situations, formation of a stable mousse is extremely rapid. The W/O emulsion formation retards the weathering processes, but weathering does occur to some extent.[7]

Further weathering of mousse leads to tar-ball formation. These consist of high molecular oil components and some water. Their physical appearance varies from brittle and hard to soft and sticky. Sizes range from millimeters to tens of centimeters. At present, tar-ball pollution is common in all oceans and is especially high in the Mediterranean and Sargasso Seas. Chemical and microbial stability of tar balls is high. Their ultimate fate in the sea consists of breakup and sinking within a period of one year.[7]

On shores, both mousse and tar balls are similarly degraded by mechanical, chemical, and biological processes, which are slowest in low energy lagoons, estuaries, and bays, where the stranded oil may persist for more than ten years.[7]

## 10.4 MAJOR OIL SPILLS AND THEIR ECOLOGICAL IMPACT

The most visible oil spills have been associated with groundings of large crude-oil-transporting tankers. The shipwrecks of the Torrey Canyon in the Channel of England in 1967, the Monte Urquiola at the coast of Spain in 1976 and the Amoco Cadiz at the coast of Brittany, France in 1978 caused severe damages to animals and long-lasting pollution of the shores.[4] The grounding of the Exxon Valdez in the Prince William Strait in Alaska (March 1989) damaged shores and wildlife of the subarctic islands and coastlines.

Also, the oil well blowouts of Ekofisk Bravo in the North Sea in 1977 and that in the Bay of Campeche, Gulf of Mexico in 1979 were spills of hundreds of thousands of tons of crude oil. Most of the shoreline damages were avoided in these cases due to weathering processes.[7]

The most dramatic ecological effects in cases of both large and small oil spills have been the disastrous effects upon seabirds. When oil contaminates feathers, the birds try to remove it by preening. Unfortunately, the oil sticks to the bill and spreads deeper into feathers. It spreads to eyes, nostrils, mouth, and then to inner organs. Oil-stained feathers are poor insulators and cannot repel water. Consequently, hypothermy, prevention of feeding, and infections follow. It has been estimated that an oil slick of two $cm^2$ on feathers of a seabird can cause death within a few days.[9]

The use of dispersants to remove oil slicks has often intensified ecological damages. In the case of the Torrey Canyon catastrophe, the toxic effects of the dispersants used on fish and mussels were much more severe than those of the oil itself. The combined effects of oil residues and dispersants have also been indicated to be fatal to corals.[11]

Since 1967, new dispersants such as synthetic polyoxyalcohols, esters, sulfonates, and phosphates have been developed for oil-spill handling. The modern dispersants are more effective but less toxic, and their ecological effects are considered to be less serious.[12] Consequently, the UK used them massively during the 1980s for oil-spill defense on the sea.[13] Mechanical collection of the surface oil and impregnation with synthetic solid foams or peat followed with bioremediation is frequently applied at present.

## 10.5 METHODS OF MONITORING

Remote sensing from aircraft is very effective in following fresh oil slicks at sea. Often they can be observed visually, because refraction of light and wave formation are different on the slick than on a clean surface. Airborne radar and infrared/UV spectrometric devices are also useful in enhancing surveys.[10]

Other survey procedures are necessary when evaporation, mousse, and tar-ball formation and other weathering effects have removed the slick. Then, the monitoring involves sampling of water, soil, sediment, and biota with proper handling (see Section 5.2). Pollution of the oceans and seas is also monitored by sampling tar balls, which are collected, e.g., by scoop nets and handled similarly to the water-oil samples.[4]

The simplest quantitation of oil in water is to weigh it after separation. In normal cases the amount of oil is too small to give any results by this procedure. Then solvent extraction and spectrometric or chromatographic determination from the extract is applicable down to very small oil concentrations. The infrared (IR) spectrum of the sample gives a qualitative fingerprint for identification of the origin of the oil pollutant.[14] Quantitation of oil in water is also possible by IR.[15]

Spectrofluorometry is a sensitive quantitation method for oil in different environmental matrixes. Two-dimensional excitation-emission spectra are also very useful for identification of the source of the spilled oil (Figure 10.5.1).[16,17]

Weathering does not affect the fluorescence spectra of crude and heavy fuel oils. Therefore, the source of the spilled oil could be identified from many types of oil samples.[16]

For chromatographic analyses of the bioaccumulated oil residues, samples are treated in a similar way as for persistent organochlorines (see Section 5.4.3). Shaking the samples with sulfuric acid, however, is replaced with alkaline digestion of the sample. As internal standards for quantitation, measured amounts of androstane (AND) and hexaethyl-benzene (HEB) are added to the sample before extraction. Hexane is used as an extraction solvent. The extract is eluted in a column of 1 g $Al_2O_3$ over 2 g of $SiO_2$ first with 6 ml of hexane (Fraction 1) and then with 10 ml of methylene chloride (Fraction 2).[5]

Fraction 1 contains aliphatic hydrocarbons, which are analyzed by GC/FID. Example of the chromatograms is shown in Figure 10.5.2. Compared to original crude oil, weathered oil and bioaccumulated oil residues contain lower amounts of n-paraffins (peaks marked C-15 to C-34), PRIS, and PHY related to

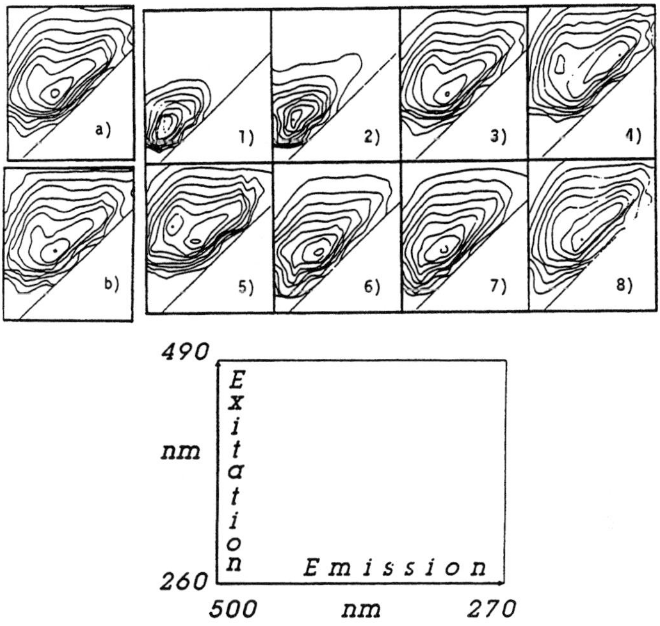


**Figure 10.5.1** Contour map of the total fluorescent spectra in a case study of a fuel oil spill. Samples: a) oiled bird feather, b) a stone, 1) to 8) various oils. Oil (3) was from the responsible tanker.[17]

branched hydrocarbons (marked as unresolved complex mixture [UCM] -hump in the chromatogram). While AND added to each sample is nearly the same, one can see from Figure 10.5.2. that mussels in the contaminated area had first collected very much aliphatics but excreted them relatively fast.

Fraction 2 consists of the nonpolar aromatic compounds that are preferably analyzed by GC/MS. Examples of GC/MS runs are given in Figures 10.5.3. and 10.5.4. The main compounds in this group are symmetrical trimethylbenzenes with a long alkyl chain (R-TMB), alkylated (methyl, dimethyl, trimethyl, tetramethyl) naphthalenes (C1N, C2N, C3N, and C4N), phenanthrenes (Ph, C1Ph) and dibenzothiophenes (especially DBT, C1DBT, and C2DBT).

Other GC detectors can be used to detect heteroaromatic components from Fraction 2. For example, flame photometrics is suitable in analysis of sulfur compounds. For nitrogen aromatics, ATD can be used.

Sulfur aromatics can be separated from aromatic hydrocar-

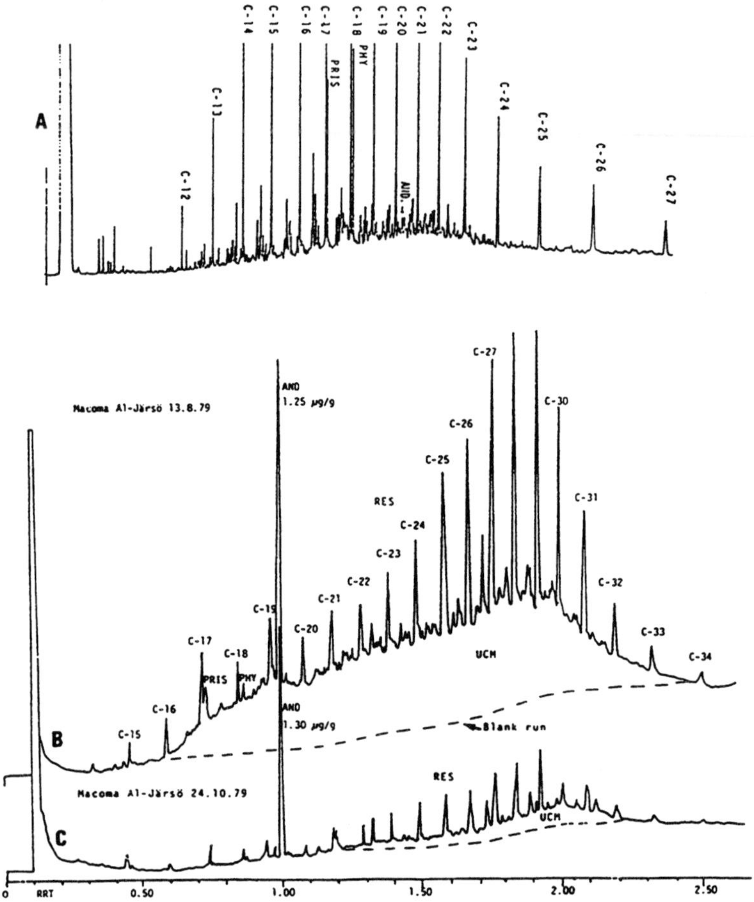


**Figure 10.5.2** FID gas chromatograms of the aliphatic hydrocarbon fractions of oil traces (Fraction 1). (A) Russian crude oil. (B) Baltic mussel (*Macoma baltica*) sample from the polluted area two months after the oil mousse was observed there. (C) Mussel sample from the same area two months later. AND represents androstane, which had been added to make an internal standard (about 1 µg/g to each sample).

bons in Fraction 2 by oxidation to sulfones, which then are separable by LC. The sulfones can be determined then by GC/FPD or by HPLC/UV.[3]

## 10.6 CASE STUDY: THE BALTIC OIL SPILL, 1979

The grounding of the tanker Antonio Gramsky off the coast of Latvia on 27 February 1979 resulted in the spill of 5000 tons

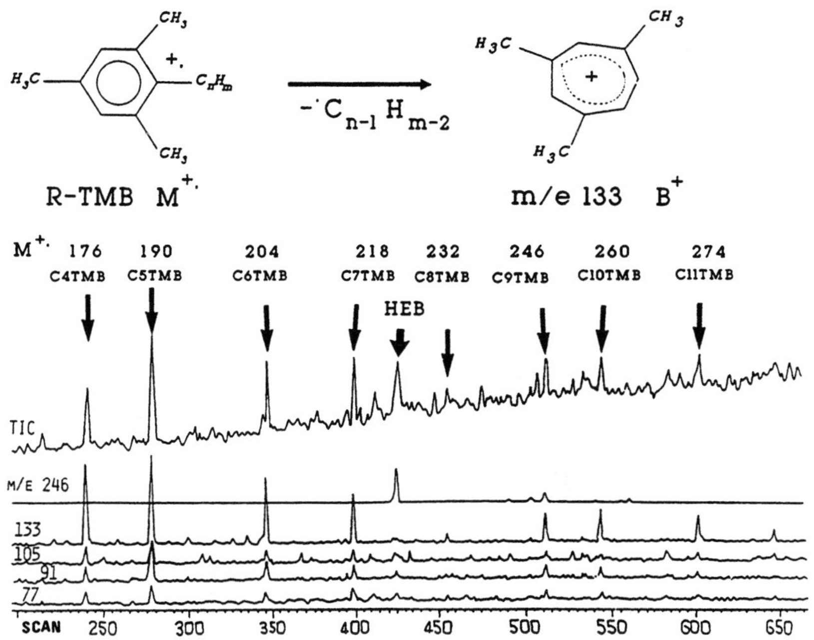


**Figure 10.5.3** GC/MS (mass chromatography mode) traces to find the main components of nonpolar oil residue aromatics from a sediment sample:[3] total ion current chromatogram (TIC) and reconstructed chromatograms of selected ions. HEB (internal standard) has m/e 246. Main R-TMB compounds are found by ions m/e 133, 105, 91, and 77.

of oil into the Baltic Sea that drifted in the northern Baltic for three months (Figure 10.6.1).

Aliphatic hydrocarbons and nonpolar aromatic compounds were analyzed from drifting weathered oil clumps, sediments and sediment trap samples, mussels, and fish at five areas of the Finnish Archipelago Sea during three years (1979 to 1981). In addition, sediments in an experimentally oil-contaminated shore area (S; see Figure 10.6.1) were analyzed annually for ten years.[3,6]

The cold winter conditions made most of the weathering processes of the drifting oil very slow. Consequently, residual oil collected from the surface had very similar aliphatic hydrocarbon ratios to a sample of Russian crude oil (Table 10.6.1).

The nonpolar aromatic compounds showed more clear weathering effects (Figure 10.6.2).

Both Macoma and Mytilus mussels collected very significant

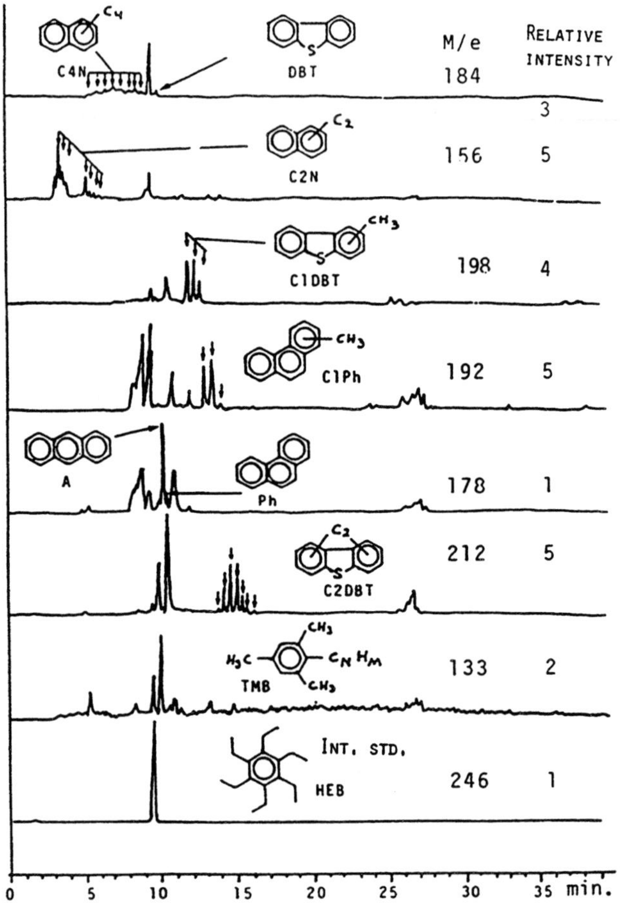


**Figure 10.5.4** GC/MS TIC and reconstructed selected ion chromatograms of nonpolar aromatics (Fraction 2) in a mussel (*Mytilus*) from an oil-polluted area of the Baltic Sea.[5]

amounts of oil residues even in areas where residual oil slicks or mousse were no more visible (Figure 10.6.3).

All oil residue levels in mussels of the polluted area decreased significantly in two months in summer 1979. The following year, in most samples, only resolved aliphatic hydrocarbons (RES) were measurable. The same results were obtained by the fish (flounder) samples.[6]

In sediment of the experimentally polluted area (S in Figure 10.6.1) most aromatic crude oil components biodegraded in three years to nondetectable levels. TMBs were detectable after

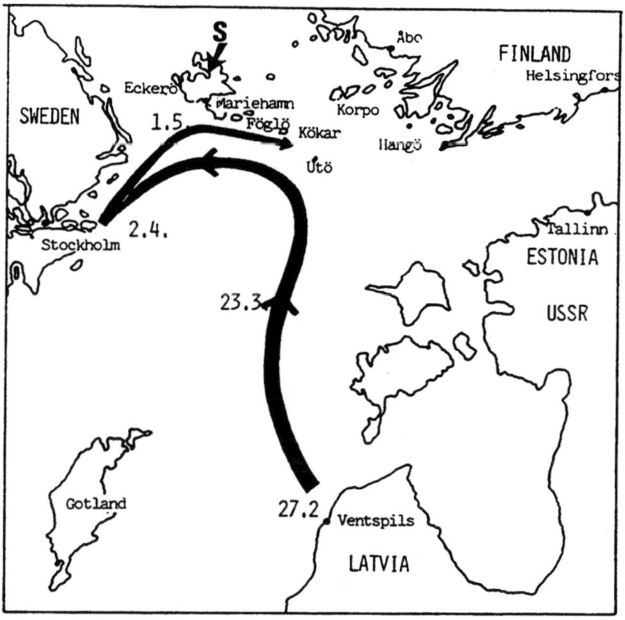


**Figure 10.6.1** Main route of the spilled oil in 1979, including the dates of the appearance of oil and the experimental site (S) that was treated with oil on October 16, 1978.[3,5,6,9]

**Table 10.6.1 Aliphatic hydrocarbon ratios in residue oil that drifted for 3.5 months in winter 1979 on the Baltic Sea, and in a USSR crude oil sample[6]**

| Sample | UCM mg/g | RES mg/g | UCM/RES | C-17/PRIS | C-17/PHY | C-17/C-18 | PRIS/PHY |
|---|---|---|---|---|---|---|---|
| Residue oil | 72.9 | 20.4 | 3.57 | 1.54 | 1.14 | 0.968 | 0.713 |
| Crude oil | 77.7 | 22.0 | 3.53 | 1.60 | 1.18 | 0.938 | 0.692 |

three years and dibenzothiophenes (DBT, C1DBT, and C2DBT) were still detectable after ten years.[3]

## 10.7 SUMMARY

Most visible ecological damage from anthropogenic uses of crude oil and oil products have taken place in association with transport and production of oil in the marine environment. Many oil spills have been fatal to seabirds. Crude oil in the sea

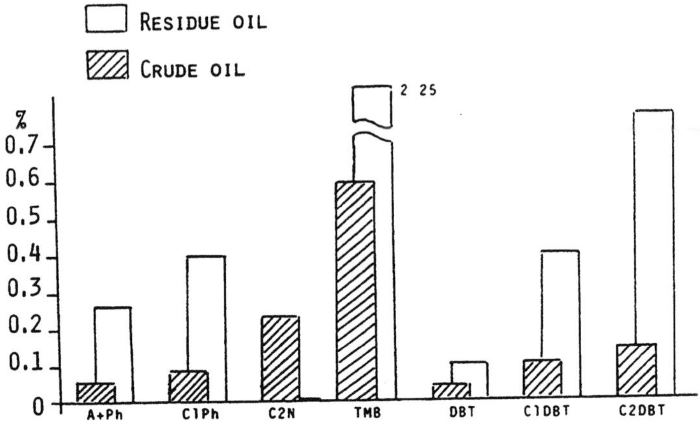


**Figure 10.6.2** Contents of the nonpolar aromatic compounds in two oil samples. A+Ph = anthracene + phenanthrene. C1Ph = methyl- phenanthrenes. C2N = dimethylnaphthalenes. TMB = sum of the alkyl-sym.trimethylbenzenes. DBT = dibenzothiophene. C1DBT = methyldibenzothiophene. C2DBT = dimethyldibenzothiophenes.[6]

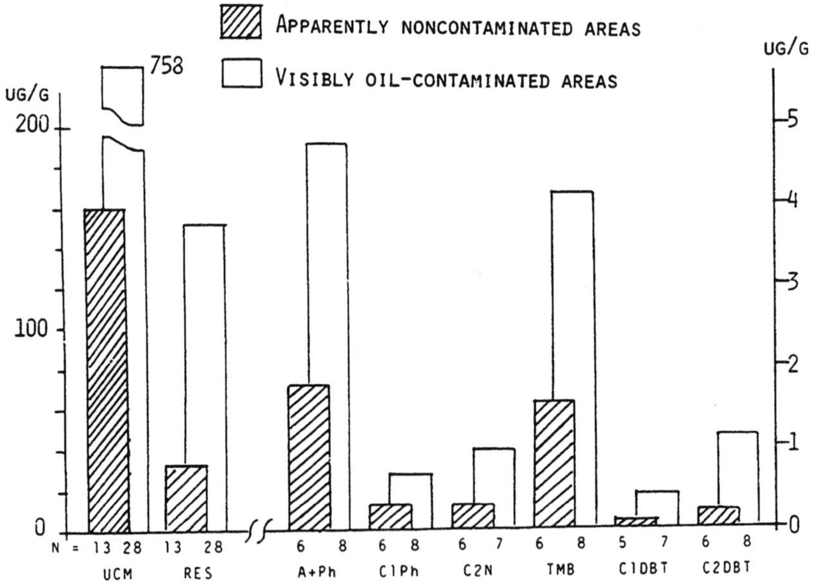


**Figure 10.6.3** Average contents of some oil residue components in *Mytilus edulis* samples collected in summer 1979 at Finnish Archipelago Sea. For name abbreviations, see Figure 10.6.2.

undergoes weathering through evaporation and transformation. Remedial activities consist of mechanical procedures and the use of chemicals. Oil trace analyses involve spectrometric and chromatographic methods. The most persistent known organic components of crude oil, aromatics, excrete from biota in few years.

## 10.8 REFERENCES


1. **Kirk-Othmer.** "Petroleum," *Encyclopedia of Chemical Technology* Vol.14 (1967) pp. 835-867.
2. **Wall, G. A.** "Petroleum, Oil Shale and Natural Asphalts," in *Materials and Technology, Vol. IV, Petroleum and Organic Chemicals* (London:Longman J.H.De Bussy, 1972) pp. 1-35.
3. **Sinkkonen, S.** "Determination of Crude Oil Alkylated Dibenzothiophenes in Environment," Department of Chemistry, University of Jyväskylä, Research Report 30(1989).
4. **Junghans, R. C., Ed.** "Marine Pollution Monitoring. (Petroleum)," NBS Special Publication 409 (1974).
5. **Paasivirta, J., R. Herzschuh, M. Lahtiperä, J. Pellinen, and S. Sinkkonen.** "Oil Residues in Baltic Sediment, Mussel and Fish. I. Development of the Analysis Methods," *Chemosphere* 10:919-928 (1981).
6. **Paasivirta, J., H. Kääriäinen, M. Lahtiperä, J. Pellinen, and S. Sinkkonen.** "Oil Residues in Baltic Sediment, Mussel and Fish. II. Study in the Finnish Archipelago 1979-81," *Chemosphere* 11:811-821 (1981).
7. **Payne, J. R., and C. R. Phillips.** *Petroleum Spills in the Marine Environment* (Chelsea, MI:Lewis, 1985).
8. **Ogata, M., Y. Miyake, K. Fujisawa, S. Kira, and Y. Yoshida.** *Bull. Environ.Contamin.Toxicol.* 25:130-135 (1980).
9. **Pfister, K., Ed.** "The 1979 Baltic Oil Spill. Environmental Studies," Ministry of the Interior, Department of Environmental Protection, Finland (1980), and other reports to Finnish authorities.
10. **Looström, B.** "Swedish Remote Sensing Systems for Oil Spill Surveillance at Sea," *Oil Petrochem. Pollut.* 1:235-241 (1983).

11. **Knap, A. H., T. D. Sleeter, R. E. Dodge, S. C. Wyers, H. R. Frith, and S. R. Smith.** "The Effect of Oil Spills and Dispersant Use on Corals," *Oil Petrochem. Pollut.* 1:157-169 (1983).

12. **Brochu, C., E. Pelletier, G. Caron, and J. E. Desnoyers.** "Dispersion of Crude Oil in Seawater: The Role of Synthetic Surfactants" *Oil Chem. Pollut.* 3:257-279 (1986/87).

13. **Turner, A. C.** "The Use of Oil-Spill Dispersants: Some Views from the UK," *Oil Chem. Pollut.* 3:417-432 (1986/87).

14. **Mattson, J. S.** "Fingerprinting of Oil by Infrared Spectrometry," *Anal.Chem.* 43:1872-1873 (1972).

15. **Carlberg, S. R.** "A Five-Year Study of the Occurrence of Non-Polar Hydrocarbons (Oil) in Baltic Waters 1970-1975," *Int.Explor.Mer.* 171:66-68 (1977).

16. **Benz, A. P.** "Oil Spill Identification," *Anal. Chem.* 48:454A-472A (1976).

17. **Mason, R. P., and G. I. H. Kerley.** "Identification of Spilled Oils by Fluorescence Spectrometry," *Oil Chem. Poll.* 4:57-70 (1988).

# Index